ELECTRONIC PROPERTIES

Structure and Properties of Materials

ELECTRONIC PROPERTIES

Robert M. Rose

Lawrence A. Shepard

John Wulff

JOHN WILEY & SONS, INC. *New York* • *London* • *Sydney*

Preface

The four brief volumes in this series were designed as a text for a two-semester introductory course in materials for engineering and science majors at the sophomore-junior level. Some curriculums provide only one semester for a materials course. We have found that under such circumstances it is convenient to use Volumes I, II, and parts of Volume III for aeronautical, chemical, civil, marine, and mechanical engineers. Similarly, parts of Volumes I, II, and IV form the basis for a single-semester course for electrical engineering and science majors.

The four volumes grew from sets of notes written for service courses during the last decade. In rewriting these for publication, we have endeavored to emphasize those principles which relate properties and behavior of different classes of materials to their structure and environment. In order to develop a coherent and logical presentation in as brief a context as possible, we have used problem sets at the end of each chapter to extend and illustrate particular aspects of the subject.

The mathematics used, for the most part, parallels that found in introductory physics, physical chemistry, and engineering textbooks. Occasional excursions above this level have only been made to bridge the gap between the present introduction and solid-state physics texts. The accent placed on real materials and engineering situations throughout all four volumes, we hope, counterbalances such lapses. Problem and laboratory sections, which supplement lectures, have proved of great help in this regard.

Many of the tables and illustrations used in the present text have been borrowed from individual specialists and their publications. Our thanks are given both to the individuals responsible for the data and to the publishers. Particular acknowledgement is given to our colleagues at M.I.T. and elsewhere who have reviewed the material presented. Many parts of the text have been improved as a result of their labors. The mistakes that remain are our own. We hope that readers who notice them will find the time to put us right.

Finally, we acknowledge our indebtedness to the Ford Foundation and to Dr. Gordon S. Brown, Dean of Engineering at M.I.T., who initially supported our efforts to provide lecture demonstrations, laboratory experiments, and notebook editions of the present text for the use of our students.

July 1965

ROBERT M. ROSE,
LAWRENCE A. SHEPARD,
JOHN WULFF, Massachusetts Institute of Technology

Contents

ELECTRONIC PROPERTIES

CHAPTER ONE

Electron Energies in Solids

If an electron is bound in any way, its energy tends to be limited to discrete levels or bands. Thus the electron bound to the nucleus of an atom can exist only in a series of states having sharply defined energies. To change its energy, the electron must jump from one state to another. In the process, it can emit or absorb a photon of electromagnetic radiation whose frequency is proportional to the energy difference between the states. The photon is viewed as a packet of energy, that is, a particle. On the other hand, a moving particle can be regarded as a packet of waves and can be assigned a wavelength which is proportional to the particle momentum. The electron energy level model can be extended to diatomic and polyatomic molecules and to solids. The key to this extension is the multiplication of levels that occurs as atoms come together to form a molecule or a solid, and the wave properties of electrons in solids. Such a model helps to explain many of the electric, thermal, magnetic, and optical properties of materials.

1.1   INTRODUCTION: QUANTA AND WAVES

Planck (1900) found that electromagnetic radiation can only be emitted or absorbed by matter in small discrete units called *photons* according to the equation

$$E_p = h\nu \tag{1.1}$$

where ν is the frequency of the radiation and h is Planck's constant (6.63×10^{-34} joule-sec). Electromagnetic radiation is, therefore, regarded as being *quantized* into particlelike *photons* of energy E_p. De Broglie (1924) postulated a converse relationship, that particles of mass m and velocity v act like waves, and the wavelength λ of the particle is

$$\lambda = \frac{h}{mv}. \tag{1.2}$$

Davisson and Germer (1927), as well as G. P. Thompson, confirmed the wave nature of electrons by showing that a beam of electrons all of the same energy could be diffracted by a crystal. Electron diffraction patterns are shown in Figure 1.1 for single crystal aluminum and polycrystalline gold.

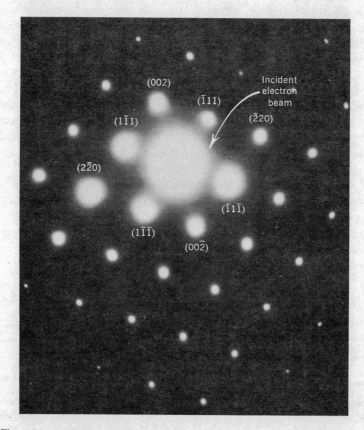

Figure 1.1 (*a*) Electron diffraction pattern from an Al single crystal; incident beam is near the [100] direction (courtesy R. D. Heidenreich).

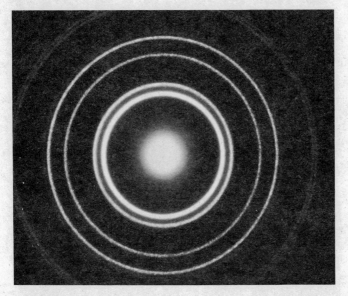

Figure 1.1   (b) A typical transmission electron diffraction pattern of polycrystalline material (gold) (courtesy J. F. Breedis).

It is important to realize that quantum mechanical relationships represent useful mathematical models. They are used to describe certain characteristics only because they predict the correct results, which is all one can ask of any model. Quantum mechanical treatments are valid on an atomic scale, but extension to macroscopic dimensions requires great care.

It is useful to describe waves in terms of the *wave number*, κ. κ is a vector quantity whose magnitude is related to the wavelength by the equation

$$|\kappa| = \frac{2\pi}{\lambda} \tag{1.3}$$

The direction of κ is the same as the velocity v. Referring to Equation 1.2, we see that the wave number is proportional to the momentum mv. The kinetic energy of a free electron which is also the total energy, according to Equations 1.2 and 1.3 is given by

$$\text{K.E.} = \tfrac{1}{2}mv^2 = \frac{h^2\kappa^2}{8\pi^2 m} \tag{1.4}$$

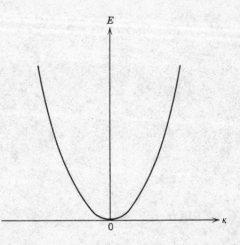

Figure 1.2   The parabolic relation between the energy and wave number of the free electron.

The relationship between E and κ is parabolic, as is shown in Figure 1.2.

1.2   ATOMIC ENERGY LEVELS

Atomic emission spectra can be excited in arcs, sparks, and with the aid of gas discharge tubes, or flames. From the measured wavelengths of spectral lines, specific frequencies can be calculated. As we stated in Chapter 1 of Volume I, the electrons in atoms and molecules exist only in well-defined energy states. Each spectral line corresponds to a change of state; when an electron changes its energy by changing state, the energy is emitted or absorbed in the form of a photon. The frequency of the photon from Equation 1.1 is

$$\nu = \frac{\Delta E}{h} \equiv \frac{E_p}{h} \qquad (1.5)$$

Thus the frequency of a spectral line may be calculated if the energy levels are known, or vice versa.

Using a specific model for the hydrogen atom, Bohr (1913) was able to show that the electron in a hydrogen atom can exist in one of a series of states whose total energy is given by

$$E_n = \frac{-2\pi^2 m_e e^4}{n^2 h^2} = -\frac{13.6}{n^2}\ eV \text{ (electron volts)} \qquad (1.6)$$

The principal number, n, is a positive integer not equal to zero. The energy, E_n, is the energy of the electron energy states; e is the charge of the electron; and m_e the electron mass. The *quantum number* $n = 1$ corresponds to the state of lowest energy. Emission occurs when the electron loses energy by changing state, and the frequency of the emitted radiation is determined by the energy lost (Equations 1.5 and 1.6). Figure 1.3, an energy level diagram for hydrogen, has arrows indicating the transitions which lead to the several frequencies, and thus (since $\nu\lambda = c$, where c is the speed of light) the wavelengths observed spectroscopically.

Bohr's calculation holds for hydrogen but not the more com-

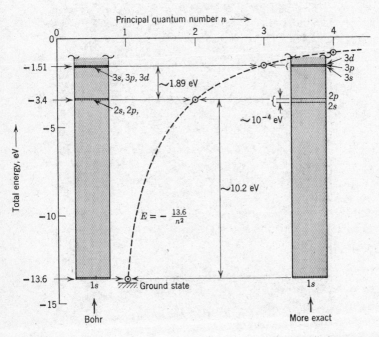

Figure 1.3   Energy level diagram for the excited states of atomic hydrogen.

plicated atoms, and more complete explanations were developed. The idea of a series of electronic states with discrete energies is nonetheless valid, although the description becomes more complex for heavier atoms, as shown in Figure 1.4 for Na. Besides the principal quantum number, n (Equation 1.6), the *azimuthal* (or "orbital") quantum number, l, is also used. The number l represents the orbital angular momentum of the atomic electron and can take on integral values from 0 to n-1. The abbreviated spectroscopic notations, s, p, d, and f, for the states $l = 0, 1, 2$, and 3, respectively, are used in the columns of the figure. To account for observed splitting of spectral lines into multiple lines when the spectrum is excited in a magnetic field, it is necessary to introduce a *magnetic* quantum number, m_l, which determines the component of the orbital angular momentum parallel to an applied magnetic field; m_l may have any integral value between $-l$ and $+l$, even when no magnetic field is present. As we shall see in

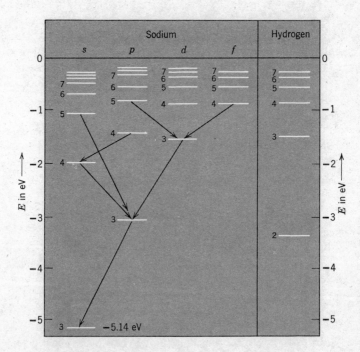

Figure 1.4   The energy level diagram for atomic sodium.

Chapters 9 to 11, there is a magnetic moment associated with the orbital angular momentum, and m_l merely specifies the orientation of that moment with regard to the field. Also it is found that electrons have intrinsic angular momentum and magnetic moments, no matter where they are; this phenomenon is called the *electron spin*. To account for the effect of spin, a fourth quantum number, m_s, is used. Only values of $+\frac{1}{2}$ or $-\frac{1}{2}$ may be assigned to m_s. Each atomic state can then be identified by a set of four quantum numbers: n, l, m_l, m_s. We now postulate the *Pauli Exclusion Principle:* each state can be occupied by no more than one electron. Otherwise we might expect all of the electrons in any atom to descend to the lowest state, $1s(n = 1, l = 0)$. In sodium, the eleven electrons occupy the two $1s$ states, the two $2s$ $(n = 2, l = 0)$ states, the six $2p$ $(n = 2, l = 1)$ states, and one of the $3s$ $(n = 3, l = 0)$ states. (See Chapter 1, Volume I and tables in the Appendix.) Because the electrons in the lower levels have very low potential energies and are so tightly bound to the core of the atom, the low thermal energy of a flame or arc usually excites only the $3s$ electron, promoting it to a higher state. The optical spectrum results from the downward transitions of this excited $3s$ electron. Since energies of the magnitude of only a few electron volts are involved, wavelengths of the order of 6000 Å are observed. Electrons nearer the nucleus can only be removed by bombardment with high-energy electrons or photons (X-rays). If they are thus removed, one of the higher-lying electrons then descends into the empty state and energy is again lost by photon emission. Since the energy transition in this case is relatively large, the emitted wavelength is but a few angstroms and lies in the X-ray region.

1.3 MOLECULAR ENERGY LEVELS

The higher the atomic number Z, the more complex is the atom, and the greater is the number of its filled energy levels. Molecules are more complicated than atoms and have a greater number of levels because the potential energy of an electron moving in the field of multiple nuclei is not simple. Also, the energies of vibration and rotation of the atom cores relative to each other are quantized in a manner similar to atomic electron energy quantizations. So many transitions are possible for an excited molecule that groups of lines in molecular spectra are called *bands*.

To illustrate how a band structure arises, we can start with widely separated atoms and bring them together to form a molecule. At infinite separation, the atoms are independent and have atomic energy levels. As the atoms draw closer together, the electrons respond to the influence of the other nuclei and electrons. The electrons gradually change states as the atoms are brought together. The Pauli Exclusion Principle then applies to the molecule as a whole. Each atomic level will in fact split into multiple molecular levels which can accommodate all the electrons which formerly were in similar single atomic levels. Thus, in a molecule of N identical atoms, each atomic level splits into N molecular levels, each of which may be occupied by pairs of electrons of opposite spins ($m_s = \pm\frac{1}{2}$). For the diatomic H_2 molecule, each level splits in two, as shown in Figure 1.5a, which applies to the 1s level. For a hypothetical six-atom, H_6 molecule, each state should split into six, as shown in Figure 1.5b, which covers both the 1s and 2s states. The 2s state splits at a larger interatomic distance than the 1s state because the 2s electrons, on

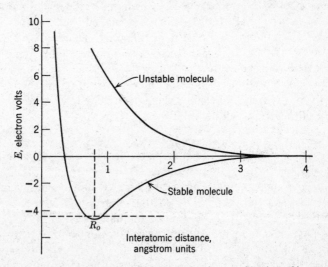

Figure 1.5   (a) The energy of a hydrogen molecule as a function of internuclear separation.

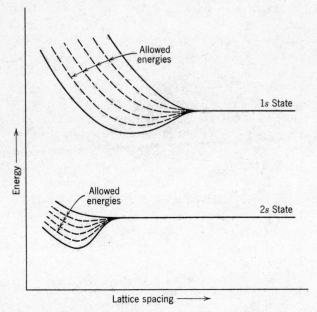

Figure 1.5   (*b*) As six atoms of hydrogen are brought together to form a linear mole-cule, the coupling between atoms increases and the energy levels split as shown.

the average, are to be found at a greater distance from the nucleus than electrons in 1*s* states.   Thus, as two atoms are brought nearer, electrons in the 2*s* states overlap, and their states begin to change before anything happens to the 1*s* states.   Such behavior is typical for atoms of larger atomic number.   The states having the largest effective radii (largest principal quantum number *n*) split first. The electron in the inner states are called *core* electrons.   Those in the outer states are called *valence* electrons, as they participate in bonding.

1.4   ENERGY BANDS IN SOLIDS

We can extend the picture of the previous section to a very large "molecule" of ordered atoms, that is, a crystalline solid.   The statement that a solid is composed of *N* atoms implies that each atomic state has split into *N* states.   Figure 1.6*a* shows the split-

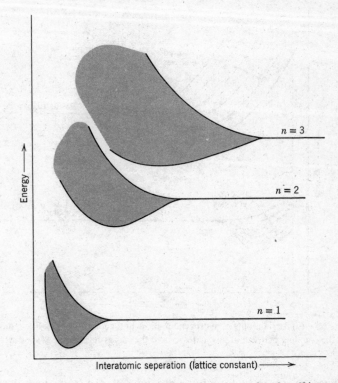

Figure 1.6 (a) The energy levels for solid hydrogen as a function of interatomic distance. The atomic levels can be considered to have broadened into bands of allowed energy. Note: the higher levels split first.

ting for solid hydrogen, and Figure 1.6b, for solid sodium. Only the envelopes of the energy levels are shown because the levels themselves are too closely spaced. The several bands in metals like sodium may overlap, but in many materials they do not. The spaces between the bands are called forbidden energy gaps, since electrons in the solid may not have these energies.

The filling of the bands follows a simple rule: states of lowest energy are filled first, the next lowest, and so on. Finally, all the electrons have been accommodated. The energy of the highest filled state is called the *Fermi level,* or the *Fermi energy,* E_F. The magnitude of E_F depends on the number of electrons per unit volume in the solid, as the latter determines how many electrons

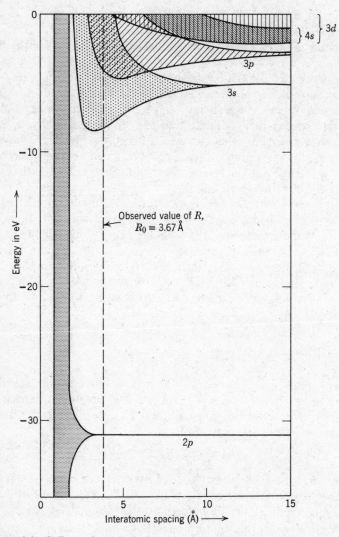

Figure 1.6 (b) Energy bands in sodium metal. According to J. C. Slater, *Phys. Rev.*, **45**, p. 794, 1934, showing splitting of the *s*, *p* and *d* levels.

must go into the bands. At $0°K$, all states up to E_F are full, and all states above E_F are empty. At higher temperatures, the random thermal energy will empty a few states below E_F by elevating a few electrons to yet higher energy states. No transitions to states below E_F will occur, since they are full. Thus, an electron cannot change its state unless enough energy is provided to take it above E_F.

Valence electrons at or above the Fermi level are, of course, in one of a number of energy states for the crystal. However, the allowable energy states in any band are so great in number and so closely spaced in magnitude that they form, essentially, a continuous range of allowed energy. A Fermi-level electron, moving through the crystal, may therefore be regarded as a *free* electron, just as the unbound electrons discussed in Section 1.1. If the electron moves with a given velocity, it will posses an associated kinetic energy and wave number, κ, in accord with Equation 1.4. Electrons in a solid will, however, be subject to certain restrictions because of the existence of the forbidden energy gaps between the bands illustrated in Figure 1.6. The continuity of the parabolic relationship between K.E. and κ, shown in Figure 1.2, must be disrupted at the energy gap because, at the gap, a large discontinuous energy is required for the electron to pass from one state to the next.

1.5   THE ZONE MODEL

Essentially the same conclusions as those of the previous section may be reached, if the wave properties of the electron are used. Being waves, electrons can be diffracted according to the Bragg diffraction law (Appendix, Volume I).

$$n\lambda = 2d \sin \theta \qquad (1.7)$$

where θ is the angle of incidence of the electron beam (or X-rays), d is the spacing between the planes of the crystal, and n is any integer. Substituting Equation 1.3 in Equation 1.7 gives

$$|\kappa| = \frac{n\pi}{d \sin \theta} \equiv \kappa \qquad (1.8)$$

According to Equation 1.8, there is at least one series of values of $|\kappa|$ corresponding to the integers n, for which electrons are diffracted and do not pass freely through the crystal. This statement should hold true whether the free electrons are part of an electron beam impinging on the crystal, or an electron in the crystal, e.g., a valence electron.   For electrons in a crystal, these values of κ correspond to the forbidden energies in the band structures. Hence, no electrons can be present which have such energies. Instead of traveling waves, Bragg diffraction leads to two sets of standing waves for the Bragg values of κ, each with different energy.   The difference between these energies corresponds to the width of the forbidden energy gap.   We therefore modify the free-electron model by interrupting the parabolic relation between κ and K.E. at the Bragg values of κ by energy gaps.   These are shown in Figure 1.7.   The energy near the forbidden region increases relatively slowly with κ and the slope actually goes to zero as κ approaches the forbidden value.   This behavior is due to increasingly strong diffraction effects as the critical value of κ is approached.

The free electrons, considered here, are the outer electrons of the individual atoms whose original atomic levels have broadened into energy bands.   The inner electrons are relatively unperturbed,

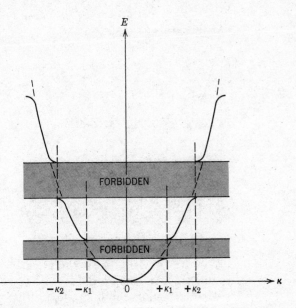

Figure 1.7   The effect of Bragg diffraction on the $E(\kappa)$ of the wavelike electron.   The slope of the curve is zero at the forbidden energy values due to diffraction effects.

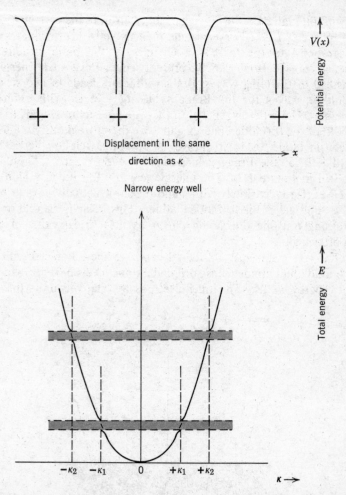

Figure 1.8 (a) The effect of ionic charge on the energy gap width. Monovalent material, singly charged, gives rise to narrow energy wells and narrow forbidden bands.

and remain close to their nuclei. The outer, or valence, electrons, which travel through the crystalline solid, see a crystal lattice of positive ions. The diffraction effects are due to the potential of the lattice, which is periodic. A deep potential well is located at each ion, due to the Coulomb force. Potential fields for singly and

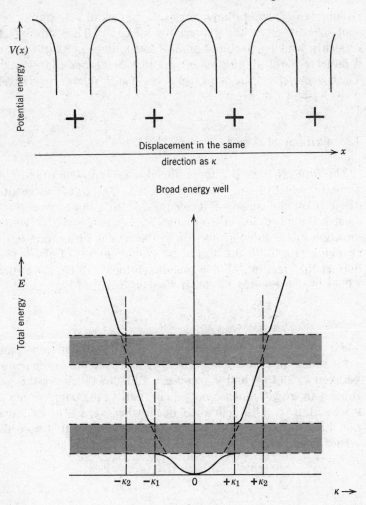

Figure 1.8 (*b*) Multiple valencies lead to wide, deep wells and wider forbidden bands.

multi-ionized atoms are shown in Figure 1.8. The size of each well, i.e., the average amplitude of the periodic potential, is a measure of the width of the energy gaps. The strength of the Bragg diffraction depends on the energy well size and stronger diffraction leads to wider gaps. The free electron, subject to a potential of zero amplitude, gives zero gap width. As the energy-well size increases, the gap widths increase. Thus multivalent elements, which give rise to multi-ionized atoms in a lattice, have a larger electrostatic attraction per ion, or deeper energy wells. The energy gaps in the band structures of such solids are relatively wide as shown in Figure 1.8.

1.6   BRILLOUIN ZONES

The *Brillouin zone* is a three dimensional representation of the allowable values of κ. In Equation 1.8, the critical value of κ depends on the angle of incidence θ. In a three-dimensional crystal, therefore, the critical value of κ depends on the relative direction of the moving electron to the crystal lattice, due to the changing value of θ, and also because different sets of planes may diffract the electron. For a one-dimensional lattice, the critical values of κ in Figures 1.7 and 1.8 would be (from Equation 1.8):

$$\kappa_n = \frac{n\pi}{a} \qquad \text{where } n = \pm1, \pm2, \pm3 \dots \qquad (1.9)$$

and a is the distance between atoms. The region between κ_1 and κ_{-1} is called the *first Brillouin zone*. The second one is the region between κ_1 and κ_2, and κ_{-1} and κ_{-2}. Figure 1.9 shows the two zones. In a two-dimensional square lattice, the components of κ are κ_x and κ_y. Diffraction will occur whenever κ satisfies Equation 1.8. If we apply Equation 1.8 and remember that both the vertical and horizontal planes can diffract, we arrive at

$$\kappa_x n_1 + \kappa_y n_2 = (\pi/a)(n_1^2 + n_2^2) \qquad (1.10)$$

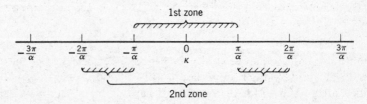

Figure 1.9   The first two Brillouin zones for a one-dimensional lattice.

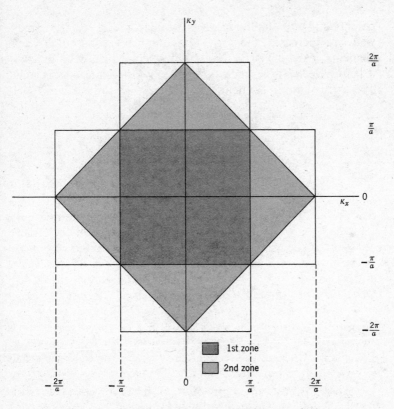

Figure 1.10   The first two Brillouin zones for a square lattice.

where n_1 is the integer for diffraction by the vertical planes, and n_2 for the horizontal planes. For the boundary of the first zone, one integer is ± 1, and the other is zero; this is the minimum value of Equation 1.10. For the boundary of the second zone, each integer is ± 1. The first two zones of the square lattice are shown in Figure 1.10. The higher zones correspond to higher values of n_1 and n_2.

The Brillouin zones of a simple cubic lattice in three dimensions can be calculated using an equation of the form

$$\kappa_x n_1 + \kappa_y n_2 + \kappa_z n_2 = (\pi/a)(n_1{}^2 + n_2{}^2 + n_3{}^2) \qquad (1.11)$$

It follows from this equation that the first zone for a simple cubic lattice is a cube (Figure 1.11) whose walls intersect the κ_x, κ_y, and

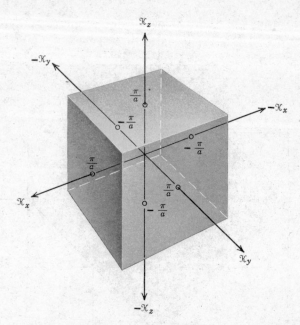

Figure 1.11   The first Brillouin zone for a cubic lattice.   The second zone consists of altitude π/a and square bases $2\pi/a$ wide resting on the cubic faces of the first zone.

κ_z axes at the points $+\pi/a$ and $-\pi/a$. The second zone in three dimensions is made up of pyramids which rest on each face of the first zone cube in a manner similar to the triangles of the second zone in Figure 1.10.

In a similar but more complicated manner, the first and second Brillouin zones for the BCC, FCC, and HCP lattices can be calculated to obtain the polyhedra shown in Figure 1.12. The equations which define the boundaries of these zones are all based on the Bragg equation for the reflection of a wave by a periodic lattice. Thus, the Brillouin zones in metals, for example, are polyhedra whose plane surfaces are parallel to the reflecting planes which are responsible for X-ray diffraction.

In both BCC and FCC metals, each Brillouin zone holds as many quantum states (without considering electron spin) as there are atoms or primitive unit cells in a crystal. For a crystal of N unit cells, there are N states in the first zone. The quantum rules apply to the filling of quantum states in a Brillouin zone. Electrons will occupy the states of lowest energy first; two electrons of opposite spin occupy each state. If there are N atoms in the crystal, then for monovalent metals the $N/2$ quantum states of lowest energy are filled. The first Brillouin zone would therefore be half filled. For multivalent metals the situation becomes more complicated because the zones may overlap. When the zones overlap it is impossible to fill up one zone without starting to fill the next one.

Besides knowing the number of quantum states in each zone, it is important to know the number of quantum states for each energy, that is, the *degeneracy* as a function of energy. The usual scheme is to calculate the number of states $N(E)$ dE per unit volume of crystal between the energy E and $E + dE$. In the coordinate system of Figures 1.11 and 1.12, (that is, κ-space), the surface of constant energy for a free electron is, according to Equation 1.4, a sphere. The volume, in κ-space, between the

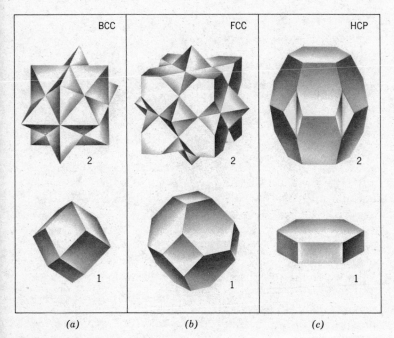

(a)   (b)   (c)

Figure 1.12   The first two Brillouin zones for the BCC lattice, the FCC lattice, and the HCP lattice.

sphere for energy E and the sphere for energy $E + dE$, contains the states between those energies. The volume per state in κ-space is $(2\pi)^3/V$, where V is the real volume of the crystal (see Problem 1.18), so the number of states between the two spherical shells is obtained by simple division. It can be shown (see references) that only positive values of κ are needed to specify all the quantum states. Therefore, we consider only the positive octant, or $\frac{1}{8}$ of the volume between the spherical shells. The density of states $N(E)$ is then (including the spin degeneracy $\pm\frac{1}{2}$) given by

$$N(E)\,dE = 4\pi V\left(\frac{2m}{h^2}\right)^{3/2} E^{1/2} dE \qquad (1.12)$$

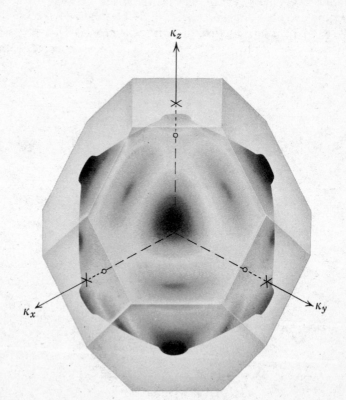

Figure 1.13   The Fermi surface of copper according to experimental measurements of Pippard. (Report on *Progress in Physics*, Vol. xxii, p. 176; *Phys. Soc.*, 1960.)

for a crystal of volume V.   Equation 1.12 is a simple parabola and applies wherever Equation 1.4 applies.   When a critical value is approached, the slope of K.E. versus κ curve approaches zero as shown in Figure 1.7, due to diffraction effects.   In three dimensions this condition means that the constant-energy surfaces touch the zone boundary at right angles.   Figure 1.13 shows just such a constant energy surface for copper (FCC).   When contact is first made with the zone boundary, a peak occurs in $N(E)$ as the sphere suddenly changes to the more complicated shape.

As expected, the lowest energy states in the zone are filled first, and then successively higher states, until all the valence electrons are accommodated.   The energy of the highest state is again called the Fermi energy, E_F.   The surface in κ-space corresponding to E_F is called the *Fermi surface*.   Figure 1.13 is actually the Fermi surface of copper, as determined experimentally by Pippard (1960).

DEFINITIONS

Electron Volt, eV.   The work done in moving an electron through a potential difference of one volt; usually abbreviated as eV; $1\ eV = 1.602 \times 10^{-19}$ joules.

Photon.   A quantized particle of radiation which is absorbed or emitted instantaneously and wholly; the photon's energy is $h\nu$, where ν is the frequency of the radiation and h is Planck's constant, 6.62517×10^{-34} joule-sec.

Free Electron.   An electron unbound by potential fields; it is able to vary its energy continuously rather than existing in the discrete energy level of atoms, molecules, or solids.

Atomic Energy Levels.   Discrete values of energy allowed to electrons which are bound to atomic nuclei by the Coulomb potential.

State.   A distinguishable quantum energy state, labeled by a characteristic set of numbers: for the atomic case, n, l, m_l, m_s.

Pauli Exclusion Principle.   Only one electron may occupy any state.

DeBroglie Relation.   Particles may behave like waves with wavelength $\lambda = h/p$, where p is the particle momentum and h is Planck's constant.

Energy Bands.   Energy levels spaced so closely together that they may be treated as continuous bands of allowed energy.

Wave Number (κ).   A vector quantity whose direction is that of propagation of the wave and whose magnitude is $2\pi/\lambda$.

Brillouin Zones.   The volumes contained in surfaces which indicate the forbidden values of κ, in κ-space.

Fermi Surface.   The surface in κ-space corresponding to E_F, the Fermi level.

Fermi Level (E_F).   The energy of the highest filled state in the highest energy band which contains electrons, in a metal, at $0°$K.

Density of States [$N(E)$].   The number of quantum states between E and $E + dE$ is $N(E)\,dE$.

BIBLIOGRAPHY

SUPPLEMENTARY READING:

Cottrell, A., *Theoretical Structural Metallurgy,* 2nd Ed., Longmans, Green & Co., New York, 1957, pp. 46–78.

Halliday, D., and Resnick, R., *Physics,* Part II, 2nd Ed., John Wiley & Sons, New York, 1960, pp. 1081–1122.

Sproull, R. L., *Modern Physics,* 2nd Ed., John Wiley & Sons, New York, 1963, pp. 1–41, 254–300.

ADVANCED READING:

Kittel, C., *Introduction to Solid State Physics,* 2nd Ed., John Wiley & Sons, New York, 1960, pp. 251–255.

Mott, N. F., and Jones, H., *The Theory of Metals and Alloys,* Clarendon Press, Oxford, 1963; Dover Press, New York, 1955, Chapters II, V.

Wilson, A. H., *The Theory of Metals,* Cambridge University Press, Cambridge, England, 1958, Chapter II.

PROBLEMS

1.1   An electron has passed through a potential difference of 5400 volts.
(a) Compute its momentum.
(b) Compute its de Broglie wavelength.
(c) Compute its wave number.
(d) Calculate the Bragg angle θ for diffraction of the electron by the (111) planes of nickel which are roughly 2.04 Å apart.

1.2   The total energy of a 1s electron in tungsten is 70,000 eV. If a 1s electron is ejected by a high velocity electron, it will be replaced by another from an outer shell (say a 2p electron). The wavelength emitted as a result of this 2p $\rightarrow$ 1s transition is about 0.21 Å. Calculate the energy of the 2p level in tungsten.

1.3   Compare Mn, Fe, Co, Ni, and Cu according to their electron configuration.

(a) How many electrons in the 3d state for each?

(b) How many electrons in the 4s state for each?

(c) How many electrons in the 4p state for each?

1.4   Distinguish between the expressions *Fermi level* and *Brillouin zone boundary*.

1.5   The copper X-ray *kα* doublet wavelength is 1.542 Å and is diffracted by the (200) planes of NaCl at $\theta = 15°51.8'$. The density of NaCl is 2.163 g/cm³.

(a) Calculate Avogadro's number. (There are four atoms per unit cell in the NaCl structure.)

(b) Calculate the value of the electronic charge. Note that one Faraday is 9.6510×10^4 coulombs.

1.6   (a) Define the ionization potential of an atom.

(b) What atoms have the highest ionization potential?

(c) How would you excite highly ionized atoms?

1.7   (a) Calculate the velocity and kinetic energy with which electrons strike the target of an X-ray tube operated at 35,000 volts.

(b) What is the minimum wavelength of the continuous spectrum emitted, assuming photons to be created in single-electron collision?

(c) What is the maximum energy per quantum of radiation if one electron produces one photon?

1.8   (a). Indicate, with the aid of a simple sketch, an electron diffraction apparatus suitable for transmission studies.

(b) Indicate with a simple sketch a suitable neutron diffraction apparatus.

(c) What are the main advantages and disadvantages to diffraction by X-rays, neutrons, and electrons?

(d) For each of these, from what do the respective beams scatter?

(*References.* Cullity, *Elements of X-ray Diffraction*, Addison-Wesley, Reading, Mass; Bacon, *Neutron Diffraction*, Oxford, New York, 1962; Harnwell and Livingood, *Experimental Atomic Physics*, McGraw-Hill, New York, 1933.)

1.9   Calculate the wavelengths of the four longest lines in the Balmer series of hydrogen. What colors do you expect them to have? (See Equation 1.6.)

1.10   X-rays from a tungsten target tube fall upon a block of silver. Using tabulated wavelengths from the *Handbook of Chemistry and Physics;*

(a) What is the smallest voltage which can be used on the X-ray tube and have the *kα₁* line emitted?

(b) What is the smallest atomic number Z of an element whose *kα₁* line is capable of exciting the *kα₁* line silver?

1.11   Suppose that an electron were confined to a box of length *l*. In the wave description, we must say that the wave function (see Volume I or the Bibliography of this chapter) must vanish at the ends of the box; that is, the nodes of the "electron waves" must always lie at the ends of the box in a manner analogous to a vibrating string clamped at the ends. This condition amounts to the stipulation

$$n\lambda = 2l, n = 1, 2, 3 \ldots$$

Restricting the problem to one dimension, derive a general expression for the electronic energy levels.

(*References*. Dicke, Wittke, *Introduction to Quantum Mechanics,* Addison-Wesley, Reading, Mass., 1960, pp. 46–55; Smith, *Wave Mechanics of Crystalline Solids,* Chapman & Hall, London, 1961, pp. 20–22.)

1.12   The "Ritz combination" law is well known in atomic spectroscopy. The law states that the sum of any two line frequencies may equal the frequency of another spectral line. Likewise, the sum of three or more may give another line frequency. Explain why. (See Equations 1.5 and 1.6.)

1.13   Compute energy levels for the hydrogen atom assuming that the electron moves in circular orbits so that the circumference at the orbit is an integral number of de Broglie wavelengths. Compare with Equation 1.6.

1.14   Consider the transition, for an atom of sodium, of an electron from the 3p state to the 3s state. From Figure 1.4, calculate approximately the frequency of the emitted radiation. Would such a frequency be emitted by solid sodium? If not, why not? (*Hint.* See Figure 1.10.) Would new frequencies be emitted? If so, explain and estimate the order of magnitude of the frequencies.

1.15   (a) The difference between an insulator and a semiconductor is supposed to be due to the difference in energy gaps. Diamond has a forbidden energy gap of 6 eV, Si 1.00 eV, Ge 0.68 eV, and Tin (grey) 0.08 eV. Illustrate the differences, using band plots.

(b) Now indicate, with a simple energy level diagram, overlapping bands in a conductor like Al.

1.16   Derive Equation 1.10.

1.17   Derive and sketch the third Brillouin zone for the square lattice. See Figure 1.10 for the first two zones.

1.18   From the Heisenberg uncertainty principle:

$$\Delta p_x \Delta x \geq h$$
$$\Delta p_y \Delta y \geq h$$
$$\Delta p_z \Delta z \geq h$$

Show that the minimum volume in κ-space per quantum state is $(2\pi)^3/V$, where V is the real volume of a three-dimensional crystal.

1.19   Derive Equation 1.12 for $N(E)$, using the procedure outlined in the text.

1.20   Derive an expression for E_F at $0°K$ in terms of the density of free electrons, for a free-electron solid.   Integrate Equation 1.12.   Explain how such an integration, from zero to E_F, should give the total number of free electrons in the solid. $[E_{F(0)} = h^2/2m(3n_0/8\pi)^{2/3}]$

1.21   Referring to Figure 1.13 and Equation 1.12, draw a density-of-states curve $N(E)$ versus E, for copper.   Indicate E_F on your drawing.

CHAPTER TWO

Electron Emission

In metals, the valence electrons are free to move and thus to conduct electricity. Thermal energy may allow some small number of electrons to escape from the surface of a metal but, in general, they are bound to it by the field of the lattice of atoms or positive ions. The energy with which they are held is known as the work function, $e\phi$. The work function of a metal may be changed by chemical and physical alteration of the surface structure. The number of electrons which leave the surface is temperature-dependent. The Richardson-Dushman equation describes the emission of electrons due to thermal energy. Electrons may also be released from a metal surface by absorption of electromagnetic radiation (photons) as in the photoelectric effect. A secondary emission of electrons occurs when an impinging beam of electrons transfers some of its kinetic energy to those in the solid. In high enough electric fields, the field at the surface of a solid may be strong enough to decrease the work function. In extremely high fields, the energy barrier for electron emission from the surface may be so narrowed that electrons can tunnel through the barrier more easily than they can surmount it.

2.1 INTRODUCTION

The obstacle to the emission of electrons from solids is the attraction of the electron to the lattice of positive ions. It is shown schematically in Figure 2.1 where the potential energy of an electron near the surface of a metal is plotted as a function of distance along a line of atom centers. Periodic potential wells are located at the sites of positive ions in this figure. The energy of electrons in the conduction (highest) band is superimposed on the

plot of potential energy versus distance.   The level for zero energy
is arbitrarily set at the bottom of the conduction band.   All the
states up to the Fermi level, E_F, are filled if the effect of temperature is not considered.   The work required to remove an electron
from the Fermi level to the vacuum outside of the metal is $e\phi$,
where $e\phi$ is the work function; ϕ is expressed in volts, and $e\phi$ in
electron volts (eV).   The voltage ϕ is required to overcome the
attraction of the positive ions at the surface of the metal.   The
magnitude of the work function $e\phi$ depends on the magnitude of
the Fermi level E_F and the height and shape of the energy barrier.
The latter depends on the composition and structure of the emitting surface.

The photoelectric effect is a convenient phenomenon for study-
ing the work function.   The effect is, in essence, the absorption of
photons by electrons in a solid, the electrons gaining enough
energy to surmount the energy barrier.   It was discovered acciden-
tally by Hertz (1887).   Although Elster and Geitel established the
basic facts about the photoelectric behavior of ordinary metals by
1889, practical application of the phenomenon in light sensors and
other devices came much later.   It required the development of

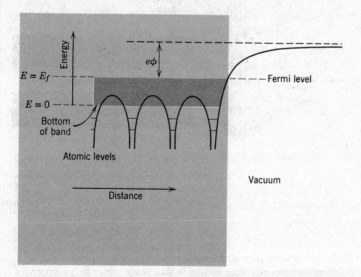

Figure 2.1   The energy barrier at the surface to the emission of electrons.

more efficient photoemitters than ordinary metals. The more rapid development of efficient thermal electron emitters may be ascribed to their earlier industrial need.

2.2   PHOTOEMISSION

The dependence of photoemission from a cathode material may be ascertained using an experimental arrangement like that shown schematically in Figure 2.2. The photoemitter is used as cathode and the silvered interior of the vacuum envelope as anode. Light of controlled wavelength and intensity is allowed to strike the cathode. When the applied anode voltage is positive, electrons are accelerated away from the cathode and the resultant photo-current is found to be independent of voltage, as shown in Figure 2.3. When the anode voltage is negative, that fraction of emitted electrons which has insufficient kinetic energy is returned to the

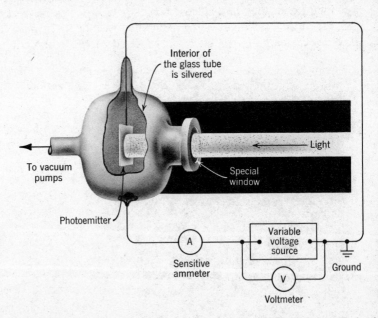

Figure 2.2   Schematic of apparatus used for the study of photoemission. Emitted electrons are collected by the silvered surface of the tube, which is connected to the voltage source.

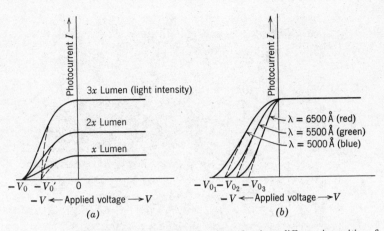

Figure 2.3   (a) Photocurrent versus applied voltage for three different intensities of light of the same wavelength; $-V_0$ at 293°K and $-V_0'$ at 0°K. (b) Photocurrent versus applied voltage for three different wavelengths; currents for positive V are arbitrarily set equal; dashed curves apply to $T = 0°$K. (According to M. Knoll and J. Eichmeier, *Technical Electronics*, Springer, New York, 1965, p. 58.)

emitter.   For an applied voltage labeled $(-V_0)$ all the emitted electrons are returned to the cathode.   This implies that the rate at which photoelectrons are emitted (the photocurrent I) depends on the light intensity, but the voltage, V_0, and therefore also the maximum kinetic energy, KE_{max} (which is equal to eV_0) do not.   If the light intensity is constant and the wavelength varied, V_0 (and the maximum kinetic energy of the emitted electrons) vary as shown in Figure 2.3(b)."

Experimental results of the kind described above led Einstein to apply Planck's quantum hypothesis (Section 1.1, Chapter One) to the problem.   In the photoelectric process, photons of energy $h\nu$ are absorbed at the solid surface as discrete units.   When such a photon is absorbed by an electron in the solid, the electron gains energy.   If the electron initially occupies the highest filled state in the conduction band and is moving normal to the emitter surface, it will leave the emitter with a maximum kinetic energy given by

$$KE_{max} = h\nu - e\phi \tag{2.1}$$

where $e\phi$ is the energy barrier for electron emission.   The left-hand side of Equation 2.1 gives the maximum kinetic energy of

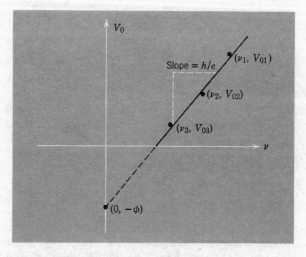

Figure 2.4   Plot of V_0 versus ν to obtain ϕ.

emitted electrons which, as mentioned previously, is equal to eV_0. We can, therefore, rewrite Equation 2.1 to obtain the Einstein photoelectric equation,

$$eV_0 = h\nu - e\phi \qquad (2.2)$$

The data of Figure 2.3b can be plotted as shown in Figure 2.4. From Equation 2.2, we see that the slope of the V_0 versus ν plot

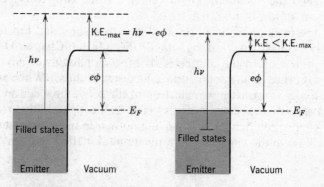

Figure 2.5   Emission of an electron (a) from the Fermi level, and (b) from below the Fermi level by a photon of frequency ν.

will be h/e, and the V_0-intercept will be equal to $-\phi$. Thus, ϕ may be measured by extrapolation. Figure 2.5 shows the excitation of electrons at and below E_F by photons of frequency ν. The kinetic energy of the emitted electrons is independent of the number of photons which strike the emitter per second. Therefore, neither KE_{max} nor V_0 varies with the light intensity, although the current I does. In summary, the kinetic energy of the emitted electrons depends only on the initial energy of the electron, the energy (or frequency) of the incident photons, and the work function.

2.3 PHOTOCATHODES

The *photoelectric efficiency* or *photoelectric yield* of a material may be defined as the number of electrons emitted per incident photon. This may also be expressed as the current emitted per unit of light intensity (amp/lumen). The photoelectric yield depends on the chemical and physical nature of the emitter as well as the wavelength of the incident radiation.

The low photoelectric yield of good conductors is due, in part, to their high optical reflectivity. It also arises from the short range of the photoelectrons produced, only about 10^{-9} meter. The depth of penetration of incident photons is 10^{-6} meter. Consequently, the layer of atoms at the surface of a metal which contributes to the photocurrent is, at the most, ten angstroms deep.

The photoelectric yield of the alkali metals, when plotted as a function of wavelength (as in Figure 2.6), indicates a rather selective response to the wavelength of the incident radiation. The yield is low at wavelengths shorter than the maximum, since here the photons penetrate the emitter to a greater depth. At wavelengths larger than that at which maximum yield is observed, reflectivity of incident photons is so high that only a small number take part in the emission process.

Although various metals coated with alkaline-earth metal films were long used as photocathodes, more efficient oxide-layer, intermetallic compound and multilayer photocathodes are used today. One efficient oxide-layer type of cathode is made by first depositing silver on a well-outgassed substrate. The silver surface is then superficially oxidized in a glow discharge. A cesium film

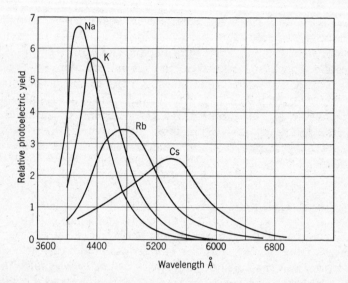

Figure 2.6   Relative photoelectric yield of the alkali metals as a function of frequency.

of controlled thickness is deposited on this surface. After heat treatment at 150°C, the surface becomes a Ag-Cs$_2$O-Cs composite of high photoelectric yield, as shown in Figure 2.7. A higher yield photocathode, whose sensitivity encompasses a rather broad spectral range is made by forming a Cs$_3$Sb film on an outgassed substrate. A thin cesium film is deposited simultaneously with

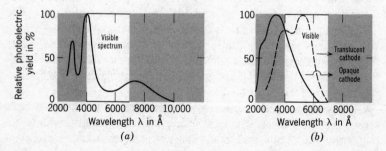

Figure 2.7   (a) Relative photoelectric yield of a Ag—Cs$_2$O—Cs photocathode. (b) Relative photoelectric yield of a Cs$_3$Sb photocathode. (According to Zworykin-Ramberg.)

antimony or following antimony. In the latter case, the compound Cs_3Sb is formed by heat treatment at 140°C. Subsequent superficial oxidation of the surface is found to increase the photoelectric yield about 50 per cent. The photoelectric yield of such cathodes in the visible region of the spectrum is even higher if the intermetallic compound film is slightly transparent. Still higher sensitivity in the visible region of the spectrum is reported for intermetallic compound photocathodes made by depositing controlled amounts of K, Na, Cs, in the order named, on top of an initially deposited Sb film. The oxide layer and the antimony compound photocathodes are really semiconductors. As such, their electrons are not as free, and the reflectivity of the cathode for incident radiation is less than that of pure metals.

2.4 THERMAL ELECTRON EMISSION

The model described in Figure 2.1 does not take temperature into account. It describes the situation at 0°K only. At higher temperatures we must consider the distribution of thermal energy among the electrons in the highest band. (We shall do this quantitatively in Chapter 3.) A certain small number have kinetic energies far above the average and well in excess of the work function. These electrons can escape from the solid. With increasing temperature, the fraction of high-energy electrons increases, and a greater number escapes, resulting in thermal electron emission.

Figure 2.8 describes a simple experimental arrangement for the investigation of thermal electron emission. The cathode (C) is heated in vacuo by the filament (F) to temperatures at which the emission is large enough to be measured. The anode, usually called the plate (P), is maintained at a positive voltage relative to the cathode. The electrons which are emitted from the cathode are then attracted to the plate, and the total emission is equal to the current flow to the plate connection. Typical thermal data are plotted in Figure 2.9.

If the emission current is high enough, the mutual repulsion of the electrons which flow through the vacuum becomes important. Since each electron experiences the electrical field of its neighbors, the repulsion of its neighbors may prevent it from reaching the plate. The so-called *space charge* becomes more important as the

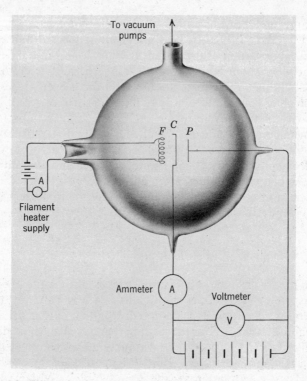

Figure 2.8   Schematic of apparatus for thermal electron emission measurement.

current density rises. The first effect of space charge is to retard the less energetic electrons. Finally, the current saturates to a constant value, as seen in Figure 2.9a. Higher cathode temperatures result in more energetic emission, and saturation occurs at higher currents as shown in Figure 2.9b. The effect of space charge may be overcome by increasing the applied voltage, as shown in Figure 2.9c. In the limit of extremely high voltage, where space charge effects are overcome, all the emitted electrons reach the anode. Nonspace charge limited thermal electron emission is described by the Richardson-Dushman equation:

$$J = AT^2 e^{-e\phi/kT} \tag{2.3}$$

where J is the current density, A is a constant for the emitting material, T is the absolute temperature, k is Boltzman's constant, and $e\phi$ is the work function of the emitter.

The value of ϕ depends on temperature according to

$$\phi = \phi_0 + \alpha T \tag{2.4}$$

where ϕ_0 denotes the value of ϕ at $T = 0°$K and $\alpha = d\phi/dT$ is the temperature coefficient of ϕ; $e\alpha \sim 10^{-4}$ $eV/°$K for most pure metals. Combining Equations 2.3 and 2.4 gives

$$J = A_0 T^2 e^{-e\alpha/k} e^{-e\phi_0/kT} \tag{2.5}$$

For pure, clean metals, the plot of $\ln J/T^2$ versus $1/T$ gives a straight line. The slope of this line is a measure of $e\phi_0/k$ and the intercept $1/T = 0$ gives the value of $A_0 e^{-e\alpha/k}$. The value of k is of the same order of magnitude as $e\alpha$ and the intercept is approximately equal to 4×10^5 amp/m$^2°$K. The value of A usually lies

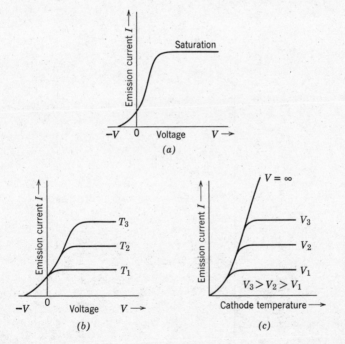

Figure 2.9   (a) Thermal electron emission current versus applied voltage for constant cathode temperature. (b) Thermal electron emission current versus applied voltage for three different cathode temperatures; cathode temperature limited. (c) Thermal electron emission current versus cathode temperature for different voltages; limited by space charge for V_1, V_2, and V_3, but not when $V = \infty$.

Table 2.1   Thermal Emission Characteristics of Different Cathodes

CATHODE FORM	MATERIAL	MELTING POINT (°K)	OPERATING TEMPERATURE (°K)	WORK FUNCTION (eV)	CONSTANT A (10^6 AMP/m^2°K^2)
Solid	W	3683	2500	4.5	0.60
Metal	Ta	3271	2300	4.1	0.4–0.6
	Mo	2873	2100	4.2	0.55
	Th	2123	1500	3.4	0.60
	Ba	983	800	2.5	0.60
	Cs	303	293	1.9	1.62
Film	W + Th	—	1900	2.6	0.03
	W + Ba	—	1000	1.6	0.15
	W — O — Ba	—	1100	1.3	0.015
Oxide-coated Metal	BaO + SrO on Ni	—	1100	1.0	10^{-4}–10^{-5}

between 3×10^5 and 7×10^5 amp/m^2°K. Table 2.1 lists the properties of three types of thermal electron emitters described in the next section.

2.5 THERMAL ELECTRON EMITTERS

Although the work function of tungsten is high (4.5 eV), its high melting point and low vapor pressure make it the most useful pure metal thermal emitter. Even so, it has major shortcomings. Unless it is made with an admix of 0.5-2 per cent of thorium oxide or lesser amounts of other additions (Al_2O_3-Na_2O-SiO_2), tungsten exhibits excessive grain growth when heated at temperatures above 2000°K with grain boundaries transverse to the wire axis. This not only leads to brittleness and loss in mechanical strength, but also to grain boundary slip and failure due to local overheating. If the proper additive (ThO_2) is employed, grain coarsening at operating temperatures (2300–2500°K) can be avoided. Specially processed wire, in which recrystallized grain boundaries are more or less parallel to the wire axes or are absent entirely (single crystal wire), is also used in some applications. In the use of such wire, proper mounting from a mechanical stress standpoint is mandatory.

Thoriated tungsten filaments, when heated in vacuo to 2500–2800°K for a short time, become better electron emitters. At this temperature a large part of the thorium oxide present in the interior is reduced to thorium metal, which can migrate to the surface by grain boundary diffusion during operation of the filament. Although the work function of bulk thorium metal is 3.4 eV, the measured work function of a tungsten filament coated with a monolayer of thorium atoms is 2.6 eV. This arises from the electric dipole layer formed at the surface, which acts as a sheet of positive charge, facilitating the escape of electrons. A monolayer of barium atoms on tungsten similarly reduces the work function of tungsten to 1.6 eV, but a monolayer of oxygen which is electronegative raises it to 9.1 eV. Nevertheless, a lesser coverage of a tungsten thermal emitter with oxygen can favor barium adsorption on the tungsten and can produce a surface whose work function is as low as 1.1 eV.

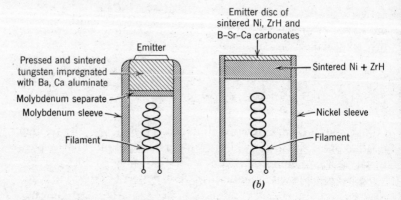

Figure 2.10   Two types of dispenser cathode.   (*a*) Sintered and impregnated tungsten emitter.   (*b*) Sintered matrix nickel emitter.

Thin film cathodes of the above kind are extremely sensitive to chemical contamination by residual gases in vacuum tubes, and removal by ion bombardment.   Where necessary, more rugged emitters, called dispenser cathodes, are used.   Such cathodes are shown in Figure 2.10.   The emitter plug in Figure 2.10*a* is made of porous tungsten impregnated with barium-calcium aluminate. The matrix plug in Figure 2.10*b* is made from a pressed and sintered disc of barium-strontium-calcium carbonate, zirconium hydride and nickel powder.   In both examples the dispenser evolves barium metal during use, which continuously forms a fresh emitting surface, and this counteracts the deleterious action of residual gas atoms or impinging ions.   Such cathodes can be effectively operated at temperatures below $\sim 1400°$K to give a high current density.

Emitter film cathodes of the kind described above should not be confused with the widely used oxide (BaO + SrO) coated cathodes introduced by Wehnelt as early as 1904.   At the present time the latter are widely used in receiving, cathode ray, television, and other vacuum tubes because they give a high emission current density at $1000°$K.   They are made by carefully coating a nickel alloy sleeve (Figure 2.11) wire or ribbon with a uniform layer of barium and strontium carbonate paste.   The paste is thermally decomposed during exhaust of the vacuum tube.   By heating the

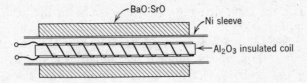

Figure 2.11   Indirectly heated nickel alloy sleeve cathode coated with 50 BaO:50 SrO.

residual oxide layer to above 1000°K for a few minutes, oxygen vacancies and free barium atoms are formed at the metal-oxide interface. The work function of such a surface is only 1.0 eV. Although the mechanism of emission from oxide-coated cathodes simulates semiconductor behavior, the exact mechanism is not fully understood.

2.6   SECONDARY EMISSION

When high-energy electrons bombard a metal surface, secondary electrons can be ejected. Secondary emission can be studied with apparatus of the kind shown schematically in Figure 2.12. The kinetic energy of the secondary electrons does not necessarily

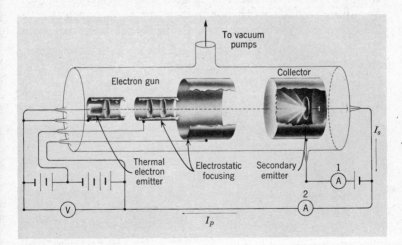

Figure 2.12   Schematic of apparatus for measuring secondary emission.

depend on the kinetic energy of the incident beam and, usually, is near 10 eV.

The efficiency of a secondary emitter is commonly measured in terms of the ratio of the number of emitted electrons I_s to the number of incident electrons I_p (see Figure 2.13). This ratio, called the secondary emission yield δ, is a function of the kinetic energy of the incident beam. If the incident beam kinetic energy is low, only a few electrons will gain sufficient energy to escape from the emitter surface. For very high kinetic energies, the incident beam penetrates the emitter before collisions occur. Since many excited secondary electrons lose their energy by other collisions before they reach the surface, they do not escape. As a consequence, the secondary emission yield δ has a maximum value as a function of the energy of the incident beam, as shown in Figure 2.13. The maximum is found to be lower in pure metals than in alloys, compounds and glasses (see Table 2.2) because more free electrons are available to effectively absorb by collision the secondary electrons generated. Secondary emission yield is also increased as the angle that the incident electrons make with the surface normal is increased.

Every effort is made to avoid secondary emission in most vacuum tube design. On the other hand, devices such as the multiplier phototube and the image-orthicon television camera tube utilize secondary emission in their operation. Such applications have led to substantial progress in the fabrication of useful devices.

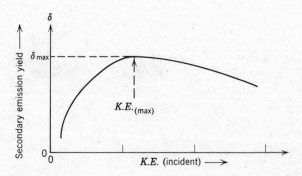

Figure 2.13   Secondary electron emission yield δ as a function of the kinetic energy of the incident electrons.

Table 2.2   Typical Secondary Emission Values

EMITTER	δ_{max}	KE_{max} KINETIC ENERGY OF INCIDENT BEAM (VOLTS)
Al	0.97	300
Cu-Mg	13.00	—
Cu	1.35	600
Cu-Al	10.00	—
Cs	0.9	400
Fe	1.32	400
Mo	1.25	375
Ni	1.3	550
Ni-B	12.0	—
W	1.43	700
MgO	8.2	525
BeO	10.2	500
Al_2O_3	4.8	1300
glass	~2.5	400
BaO-SrO	10.0	1400

2.7   SCHOTTKY EFFECT

As mentioned previously, increasing the electric field between emitter and plate can eliminate space charge effects. High fields also increase emission by altering the potential-energy barrier at the cathode surface. This is shown schematically in Figure 2.14a. When the accelerating electric field $\mathcal{E}$ is applied, it contributes to the potential energy an amount $-e\mathcal{E}x$ at a distance x from the surface. The potential energy barrier has a maximum value at the distance x_0, and the effective work function is reduced by the amount $e\Delta\phi$, as shown in the figure.

2.8   ELECTRON FIELD EMISSION

When an electric field is applied to a cathode surface and the work function is reduced, the potential barrier at the surface is also narrowed, as shown in Figure 2.14b. When the barrier be-

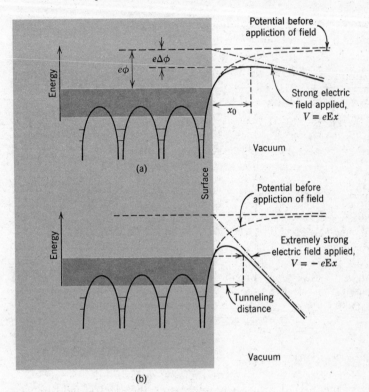

Figure 2.14   Energy diagram for electron emission in strong electric fields: (a) Strong fields decrease the apparent energy barrier for electron emission, the Schottky effect. (b) Very strong fields narrow the energy barrier and allow quantum-mechanical tunneling.

comes thin enough (~ 100 Å), electrons should penetrate it, even if they have insufficient energy to surmount it. This phenomenon is called quantum mechanical *tunneling.* Tunneling becomes appreciable when the applied field is about 10^{10} volts/meter.

The field emission current depends on the applied field in a manner analogous to the dependence of the thermal electron emission on temperature (Equation 2.3). The electron field emission is used to study surface phenomena such as adsorption, and variation of the work function with crystallographic orientation. The necessary field strength is obtained by using high applied voltages and sharp pointed wire emitters. If the applied field is reversed, gas atoms are ionized at the points and positive ions rather than electrons are pulled away from the surface. This is called ion field emission. The resolution of the ion emission pattern is so great that it can magnify individual atoms at the surface. Figure 2.15 is a schematic representation of a typical apparatus used for field-ion microscopy. Figure 2.16 shows an ion micrograph of a platinum needle. The center of the picture is the [001] direction. Each light spot corresponds to an individual atom. The four $<111>$ directions are above, below, and to the sides of the [001], and the $<110>$ directions run out at 45° angles. The atomic packing in the planes is reproduced in this micrograph.

Because extremely large amounts of power may be handled by a tiny field-emitting needle, many potential applications exist. Only one is now in use, namely, ultrahigh-speed X-ray photography. Others await the development of emitters which can yield higher currents at lower voltages.

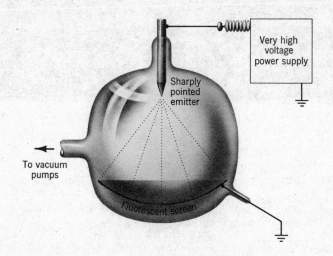

Figure 2.15 Schematic of field-emission microscope. The emission pattern is viewed on the fluorescent screen. The voltage is of the order of 10^5 volts. For ion micrography a small quantity of inert gas is released into the vacuum.

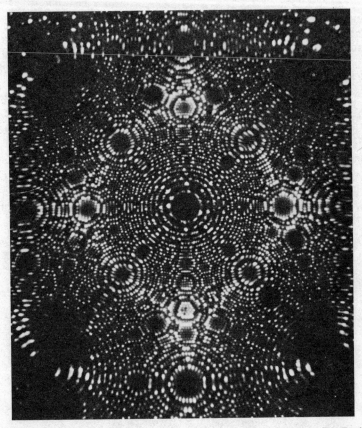

Figure 2.16 Field ion micrograph of a platinum needle. (Courtesy Dr. Erwin Müller.)

DEFINITIONS

Fermi Level.   The highest filled electronic energy level in a solid at $0°K$. It is constant for a particular metal.

Work Function $e\phi$.   The energy $e\phi$ is the energy difference in electron volts between the free state and the highest full state in the valence band (Fermi level).

Thermal Electron Emission (also called thermionic emission).   Electron emission by thermal excitation of electrons from the Fermi level to the free state.

Space Charge.   The field of emitted electrons, at high current densities, must be added to the other fields present.   The observed effects of such fields which reduce emission are called space charge effects.

Photoemission.   Electron emission from a solid by excitation of electrons by radiation.

Spectral Sensitivity.   In a photoemitter, the number of photoelectrons emitted per incident photon.

Secondary Emission.   Electron emission due to collisions of primary electrons with the conduction electrons of the solid.

Schottky Effect.   Apparent reduction of the work function due to high applied electric fields at the surface.

Field Emission.   Direct quantum mechanical tunneling of the electrons *through* the energy barrier which is narrowed by the application of extremely high electric fields at the surface of the metal.   No energy is required to excite the electrons across the work function barrier.

BIBLIOGRAPHY

SUPPLEMENTARY READING:

Azaroff, L. V., and Brophy, J. J., *Electronic Processes in Materials,* McGraw-Hill Book Co., New York, 1963.

Frank, N. H., *Introduction to Electricity and Optics,* McGraw-Hill Book Co., New York, 1950, pp. 219–225.

Kohl, W. H., *Electron Tubes,* Reinhold Corp., New York, 1960.

Millman, J., and Seely, S., *Electronics,* 2nd Ed., McGraw-Hill Book Co., New York, 1951.

Sproull, R. L., *Modern Physics,* 2nd Ed., John Wiley & Sons, New York, 1964, Chapter 12.

Stanley, J. K., *Electrical and Magnetic Properties of Metals,* A. S. M., Metals Park, Ohio, 1963, Chapter 3.

ADVANCED READING:

Bruning, H., *Physics and Applications of Secondary Electron Emission,* Pergamon Press, London, 1954.

Herring, C., and Nicholas, M. H., "Thermionic Emission," *Reviews of Modern Physics,* Vol. 21, p. 185, 1949.

Zworykin, V. K., and Ramberg, E. C., *Photoelectricity and Its Applications,* John Wiley & Sons, New York, 1949.

PROBLEMS

2.1   What temperature produces the thermal energy, kT, of one electron volt?

2.2 What is the minimum frequency necessary to achieve photoemission from tantalum? (See Table 2.1.) In choosing a photoemitter, what considerations other than the work function are important? Having chosen the material, how could you improve the emission current?

2.3 For a cesium-coated oxide-silver sandwich photoemitter, $e\phi = 0.9$ eV:

(a) What is the longest wavelength which will eject photoelectrons?

(b) If light of 3200 Å wavelength is incident on the emitter, what is maximum velocity of the ejected electrons?

2.4 The photoelectric threshhold is the maximum wavelength of incident light that will produce photoemission. Photoelectric threshholds for polished or clean metal surfaces are: for Al, 4700 Å; for Cu, 3000 Å; for W, 2300 Å; for Na, 5400 Å. Determine the work function, in eV, remembering $c = \lambda\nu$, where c is the speed of light.

2.5 The threshhold wavelength for a sodium surface is 5400 Å.

(a) Calculate the work function ($e\phi$) for sodium.

(b) What is the maximum kinetic energy of a photoelectron from the surface when the impinging light has a wavelength 2000 Å?

2.6 (a) Explain why the photoelectric emission of photocathodes should decrease when a monolayer of oxygen atoms forms on its surface.

(b) Why might a monolayer of hydrogen atoms increase photoemission?

2.7 For the Richardson-Dushman equation, the value of A for tungsten is 0.60×10^6 amp/meter^{2}deg^2, and the work function is 4.5 eV.

(a) For a cathode 0.010 in. in diameter and 1 in. long, what is the maximum emission current possible at 2000°K? At 2500°K? What is the ratio of these two currents?

(b) If the cathode is thoriated tungsten, $e\phi = 2.7$ eV. To obtain the same emission as the ordinary tungsten cathode at 2500°K, at what temperature, approximately, should the thoriated tungsten cathode be operated?

(c) For a cesium-coated tungsten cathode, $e\phi = 1.36$ eV, $A = 0.032 \times 10^6$ amp/m^{2}deg^2. What temperature is now necessary to get the same emission as part b?

2.8 The power radiated, per unit area, of a cathode in empty space is $5.7 \times 10^{-8}T^4$ watts/m^2, where T is the absolute temperature. Consider the efficiency of the emitter as:

$$E = \frac{\text{(current emitted thermally)}}{\text{(power radiated)}}$$

Using Equation 2.1, obtain an expression for the temperature at which E is maximized. For tungsten ($e\phi = 4.5$ eV, $A = 0.60 \times 10^6$ amp/m^2 deg^2), what temperature is this? At what temperature would you operate a tungsten cathode?

2.9   In the manufacture of oxide-coated nickel cathodes, the alloy content of the nickel is important.   Explain this, using equations showing the reactions that take place at the alloy interface.

2.10   With the aid of simple sketches, show how you could avoid tungsten contamination from the electron gun used in electron-beam zone refining or melting of refractory metals like Mo, Ta and Nb.

2.11   Describe how you would measure the rate of evaporation of barium from an oxide-coated cathode using the radioactive tracer technique.

2.12   With the aid of simple sketches, describe some typical ion sources of the gas type with and without the use of thermal electron emitters.

2.13   Indirectly heated dispenser anodes of the Kunsman type can be used to produce a relatively weak ion current in high vacuum.   They are made by using a mix of nickel powder, iron oxide, aluminum oxide, and a small percentage of alkali compound which is pressed and sintered to form the dispenser.   Describe how they function.

2.14   (a) Using a simple sketch, describe the construction and operation of an X-ray tube with a molybdenum target.

(b) Describe the three ways in which the electrons that strike the anode lose their energy.

(c) What X-ray wavelengths are emitted when the potential difference between anode and cathode is 35 KV?   Use a plot of intensity versus wavelength.

2.15   Describe a typical radioactive source of electrons with minimum γ and α emission.

2.16   Tungsten and molybdenum, unlike most other metals, become brittle at room temperature after recrystallization.

(a) Explain the cause with the aid of simple sketches.

(b) When ordinary wire is heated for a long time at $\sim 2600°$ K it develops a bamboo structure.   Illustrate this with a simple sketch.

(c) How would a filament with such a structure fail at elevated temperatures?

(d) Show with a sketch what type of grain boundary structure is preferred.   Describe how it is achieved.

2.17   In the fabrication of oxide-coated cathode wire, ribbon, and sleeves, it is important to coat the metal with a uniform layer of carbonate paste.   Show how this can be accomplished with the aid of simple sketches.

2.18   The oxide-coated thermal electron emitter is often referred to as a semiconductor.   Explain.

2.19   In a special type of photomultiplier tube, electrons are ejected by radiation from a Cs_2Sb photoemitter ($e\phi = 1.8\ eV$); the photoelectrons then strike a cesium surface ($e\phi = 1.9\ eV$). The secondary electrons are collected on another electrode.

(a) What is the longest wavelength that will give a secondary electron current?

(b) Referring to Table 2.2, at what wavelength will the secondary current be maximized, ignoring the spectral sensitivity variation of the photo-emitter?

2.20   Referring to page 464 of the textbook by Sproull (see the Bibliography of the chapter), describe the operation of a photomultiplier tube. The gain of a multiple-dynode tube in R^n, where R is the gain of a single dynode, and n is the number of dynodes. Relate R to δ. What total gain would you expect from a 16-dynode tube if the best material from Table 2.2 were used?

2.21   By a sketch, indicate where the $<100>$, $<110>$, $<111>$, and $<112>$ directions are in the platinum needle of Figure 2.16. What is the approximate magnification of Figure 2.16? Show that the picture reveals the atomic packing on the close-packed planes.

CHAPTER THREE

Thermal Behavior

The two principal types of thermal energy in most solids are the vibrational energy of the atoms about their mean lattice positions and the kinetic energy of the free electrons. As a solid absorbs heat, its temperature rises and its internal energy is increased. Thus, the important thermal properties, heat capacity, thermal expansion, and thermal conductivity are dependent upon energy changes of the atoms and free electrons.

The thermal energy present as lattice vibrations is looked upon theoretically as a series of superimposed sound or strain waves with a frequency spectrum determined by the elastic properties of the crystal. A quantum of elastic energy is called a *phonon*.

The heat capacity of solids is zero at $0°K$, and rises rapidly as temperature rises to an approximately constant value at higher temperatures. The increase in heat capacity corresponds to an increased capability of the phonons and electrons to increase their average energy. Because of the exclusion principle, the ability of an electron to increase its energy depends on the availability of empty states at higher energies. Generally, only electrons near the Fermi level have access to empty states, and the electronic contribution to the specific heat is relatively small.

Thermal expansion in solids arises from the asymmetry of bonding forces between atoms. As a lower force is required to separate the atoms in a crystal than to bring them closer together, increased thermal vibration naturally tends to increase the average atomic spacing.

Thermal conductivity is the transfer of heat through a solid by the phonons and electrons. Metals which are the best thermal conductors conduct heat mainly by the free electrons.

Table 3.1   Conversion Factors for Thermodynamic Constants and Units[a]

	cgs	mks
Boltzmann's constant (k)	1.38×10^{-16} erg/°K	1.38×10^{-23} joule/°K
Avagado's number (N)	6.025×10^{23} molecule/g mole	6.025×10^{26} molecule/kg mole
Gas constant (R)	1.987 cal/mole °K	8.31×10^3 joule/kg mole °K
Planck's constant (h)	6.62×10^{-27} erg-sec	6.62×10^{-34} joule/sec
Electron volt (eV)	1.60×10^{-12} erg	1.60×10^{-19} joule
Electron charge (e)	1.602×10^{-20} emu 4.80×10^{-10} esu	1.602×10^{-19} coulomb

[a] 1 joule = 10^7 erg = 0.2389 gram-calorie.

3.1   INTRODUCTION

The thermodynamic concept of heat capacity measured at constant volume or pressure was introduced in Volume II, Chapter 1:

$$C_v = \left(\frac{dE}{dT}\right)_v \qquad C_p = \left(\frac{dH}{dT}\right)_p \qquad (3.1)$$

where E is internal energy and enthalpy $H = E + PV$. In the present chapter, we shall discuss this and other thermal properties of solids from an atomistic viewpoint.*

Specific heat is defined as heat capacity per unit mass and is designated by a lower-case c_v and c_p with units of calories per mole per degree Kelvin. The classical theory of specific heat has as its foundation the law of Dulong and Petit (1819): the specific heat is the same for all elementary solid substances, about 6 cal/mole °K. Fifty years later, Boltzman pointed out that this result could be rationalized in terms of the energy of vibrating atoms in the solid. From the kinetic theory of gases, the average kinetic energy of a moving particle resolved along one coordinate direction is

---

* Relevent thermodynamic constants and units of measure are given in cgs units in accordance with conventional usage. A conversion chart from cgs to MKS systems appears in Table 3.1.

$$KE_x = \tfrac{1}{2}kT \tag{3.2a}$$

Thus the average kinetic energy per particle in three dimensions is

$$KE_{xyz} = \tfrac{3}{2}kT \tag{3.2b}$$

and the energy of a mole (N) of moving particles is

$$KE_{\text{mole}} = \tfrac{3}{2}NkT = \tfrac{3}{2}RT \tag{3.3}$$

the internal energy of a perfect gas. For bound atoms, vibrating in a solid, there are three additional components of potential energy, each equal, on the average, to $\tfrac{1}{2}kT$. Therefore, the internal energy per mole of solid is

$$E = 3RT\frac{\text{cal}}{\text{mole}} \tag{3.4}$$

and

$$c_v = \left(\frac{dE}{dT}\right)_v = 3R = 5.96\ \frac{\text{cal}}{\text{mole }°\text{K}} \tag{3.5}$$

the value obtained by Dulong and Petit.

The Dulong and Petit law is, generally, correct at room temperature and above for elements of atomic weight greater than 40. At very low temperatures, the specific heats of all elements approach zero. A number of light, high melting point elements (B, Be, C, and Si, for example) show very much lower room temperature values than the Dulong and Petit law predicts. Finally, for a number of very electropositive metals (Na, Cs, Ca, and Mg), c_v rises with increasing temperature considerably above $3R$.

Einstein (1907) dealt with the first two of these observations successfully by the application of quantum theory to the specific heat problem, assuming that the atoms vibrate independently. Although the theory correctly predicts the shape of the c_v-T curve, it shows considerable deviation from measured values at low temperatures. Debye and others demonstrated that the error lay in Einstein's assumption of independently oscillating particles. Debye took into account the interaction between vibrating particles by hypothesizing a spectrum of elastic standing waves traversing the crystal, somewhat like a bowl of vibrating jelly, that approximates the thermal vibration of atoms. The c_v-T curve determined from this model proved such a satisfactory representation that for many years the Debye extrapolation was used to provide standard values at very low temperatures.

The final observation, the unusually high specific heats of the electropositive elements, suggested that the free electrons may also contribute to the specific heat. According to classical theory, the kinetic energy of N free electrons in the crystal should be, from Equation 3.3, $\frac{3}{2}RT$. Therefore, the electronic contribution to the specific heat should be $\frac{3}{2}R$. In actuality, the electron contribution in normal solids is extremely small, and only measurable at very low and very high temperatures.

The relatively small contribution of the kinetic energy of free electrons to the specific heat is due to the fact that almost all the electrons lie in energy states which are surrounded by filled states. The Pauli Exclusion Principle forbids transitions to filled states. Empty electron states are largely available only above the Fermi level, and a relatively small fraction of the total number of free electrons are excited thermally to these states, even at high temperatures because the electron must lie within kT of E_F.

3.2   THE LATTICE SPECIFIC HEAT

The work of Dewar in the early part of this century clearly showed that the law of Dulong and Petit failed at low temperatures as shown in Figure 3.1. Einstein was the first to resolve the problem by the application of the quantum theory together with

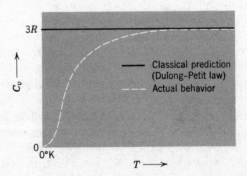

Figure 3.1   The classical prediction and actual variation of the lattice specific heat with temperature.

the assumption that the crystalline solid is made up of N atoms vibrating in the three independent directions with a constant frequency ν_E. Following Planck's quantum hypothesis, each of the $3N$ harmonic oscillators can have the quantized energies:*

$$E_n = nh\nu_E \tag{3.6}$$

where the quantum number $n = 0, 1, 2 \dots$. The number of oscillators N_n in each energy state relative to the number in the zero energy state N_0 can be determined from the Boltzmann function (Volume II, Chapter 4):

$$N_n = N_0 e^{-(E_n/kT)} = N_0 e^{-(nh\nu_E/kT)} \tag{3.7}$$

The average energy of an oscillator is (see Problem 3.3):

$$\bar{E} = \frac{E}{N} = \Sigma_n \frac{(N_n E_n)}{\Sigma_n N_n} = \frac{h\nu_E}{e^{(h\nu_E/kT)} - 1} \tag{3.8}*$$

Equations 3.6 and 3.8 are the original equations with which Planck initiated the quantum theory.

Considering $3N$ independent oscillators (N atoms, three dimensions), we find that

$$E = \frac{3Nh\nu_E}{e^{(h\nu_E/kT)} - 1} \tag{3.9}$$

and the Einstein specific heat,

$$c_v = \left(\frac{dE}{dT}\right)_v = 3Nk\left(\frac{h\nu_E}{kT}\right)^2 \left[\frac{e^{h\nu_E/kT}}{(e^{h\nu_E/kT} - 1)^2}\right] \tag{3.10}$$

The Einstein specific heat curve is a fairly good representation of the observed specific heat, Figure 3.1. For one mole, $Nk = R$, we can see that Equation 3.10 differs from the classical c_v, Equation 3.5, in the temperature dependent terms in the brackets. At high temperatures, the Einstein specific heat approaches a value of $3R$, and at low temperatures, zero (see Problem 3.4). The value of ν_E, the Einstein frequency, is usually determined by fitting the curve to experimental data. However, at low temperatures, the Einstein c_v approaches zero much more rapidly than the measured values. The source of error lies in the assumption that all of the

---

*Actually a "zero point energy" of $\frac{1}{2}h\nu$ should be added to Equations 3.6 and 3.8. It will not, however, affect the calculation of c_v.

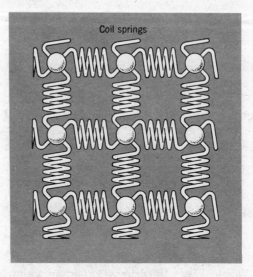

Coil springs

Figure 3.2   Elastic coupling of vibrating atoms in a lattice.

atomic oscillators vibrate independently at the same frequency. They should, in fact, be coupled as shown in Figure 3.2. This requires a rather complicated mathematical analysis, since a large number of particles are involved. The goal of the analysis is to find the frequency spectrum $g(\nu)$, where $g(\nu)\,d\nu$ is defined as the number of allowed frequencies of vibration (or, equivalently, the number of oscillators having these frequencies) between ν and $\nu + d\nu$.

Debye simplified the problem by considering a solid as a continuous vibrating medium. The distribution of lattice frequencies based on this approximation is

$$g(\nu) = \frac{4\pi\nu^2}{c_s{}^3} \qquad (3.11)$$

where c_s is the velocity of sound in the solid. Debye also postulated a maximum frequency of oscillation, ν_D, because the total number of allowed frequencies should not exceed $3N$ (N atoms vibrating in three dimensions). Also, the minimum wavelength $\lambda_D = c_s/\nu_D$ should not be shorter than the spacing between atoms in the crystal. Integrating $g(\nu)\,d\nu$ times the average energy of an oscillator, Equation 3.8, he obtained for a molar volume of crystal,

$$E = \frac{9N}{\nu_D{}^3} \int_0^{\nu_0} \frac{h\nu}{e^{h\nu/kT}-1} \nu^2 \, d\nu \qquad (3.12)$$

If we define $h\nu_D/kT \equiv \theta_D/T$ (θ_D is called the *Debye temperature*), then the Debye specific heat is

$$c_v = \frac{dE}{dT}\bigg]_v = 9Nk\left[\left(\frac{T}{\theta_D}\right)^3 \int_0^{\theta_D/T} \frac{e^x x^4 \, dx}{(e^x-1)^2}\right] \qquad (3.13a)$$

or

$$c_v = 3NkD(\theta_D/T) \qquad (3.13b)$$

where $D(\theta_D/T)$ is defined as three times the expression in square brackets in Equation 3.13, and is called the *Debye Function*. Although the Debye function cannot be integrated analytically, it approaches the limiting values:

$$D(\theta_D/T) \to 1 \qquad \text{when} \quad T \to \infty \qquad (3.14a)$$

$$D(\theta_D/T) \to \frac{4\pi^2}{5}\left(\frac{T}{\theta_D}\right)^3 \qquad \text{when} \quad T \ll \theta_D \qquad (3.14b)$$

Thus, c_v approaches the classical value $3R$ at high temperatures. At low temperatures:

$$c_v = \frac{12\pi^4}{5} R\left(\frac{T}{\theta_D}\right)^3 = 464.5\left(\frac{T}{\theta_D}\right)^3 \qquad (3.15)$$

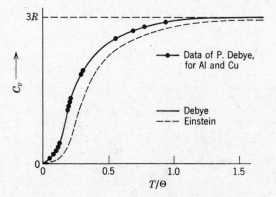

Figure 3.3 Calculation of the lattice specific heat by Einstein and Debye models, with data for Al and Cu. The data is from the work of P. Debye, *Annalen der Physik*, Vol. 39, p. 789 (1912).

Figure 3.3 shows a typical Debye c_v plot along with the Einstein c_v and the measured data. The specific heat for all substances following the Debye equation fall on a single curve when c_v is plotted against T/θ_D. Plotting log c_v versus log T for low-temperature data for various materials will give θ_D directly, as the displacement from the standard curve will be $(-3 \log \theta_D)$.

Debye viewed the thermal vibration of atoms in a solid as a mixture of phonons because the transport of sound in solids is an elastic wave phenomenon. The Debye spectrum (Equation 3.11) is often called a phonon spectrum. A phonon, as mentioned previously, is a quantum of elastic energy analogous to the photon, which is a quantum of electromagnetic energy.

3.3   ELECTRONIC SPECIFIC HEAT

As pointed out in Chapter 2, even at absolute zero of temperature, the electronic energy levels in a solid are filled right up to the Fermi level E_F. This electronic energy cannot be used to heat a cold object. Energy can only be given off by an electron transition to a lower state and, since all the lower states are filled, this is impossible. At $0°$K, only the states which have energies greater than E_F are unfilled. In Figure 3.4, the probability $F(E)$ of finding a given energy level filled is plotted against energy E.

At higher temperatures, an electron may acquire thermal energy of the order of kT, and proceed to a higher energy state, provided

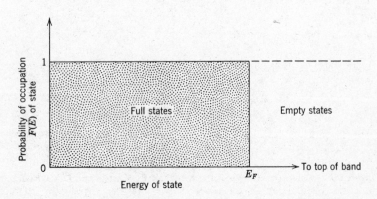

Figure 3.4   Filling of states in a partially full energy band at $0°$K.

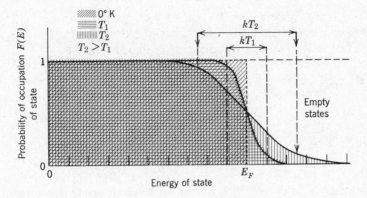

Figure 3.5   The Fermi-Dirac distribution for filling of states as a function of temperature.

that the state is not occupied.   Figure 3.5 shows the distribution of electrons at absolute zero and two higher temperatures.   We see that only electrons between E_F and $(E_F{-}kT)$ are likely to be raised to the empty states above E_F.   In both Figures 3.4 and 3.5, the probability $F(E)$ that a state is occupied is called the *Fermi function*.   It is equivalent to the Maxwell-Botzman distribution, but is derived for electrons using Fermi-Dirac statistics (see Kittel, p. 251) to give

$$F(E) = \frac{1}{1 + e^{(E-E_F)/kT}} \tag{3.16}$$

For $E \ll E_F$, the exponential term in the denominator is very small and $F(E)$ approaches one, i.e., the particular state is most probably filled.   Only electrons near the Fermi level are influenced by increasing the temperature and it is these that contribute to the specific heat.   An electron at E_F will, on the average rise to $(E_F + kT)$, and most likely it will end up at a somewhat lower energy.   Electrons having energies much less than $(E_F - kT)$ do not contribute very much to the electronic heat capacity of the solid, since almost all the states within kT above them are full, and adding another electron violates the exclusion principle.   Of all the electrons present, only the fraction of the order of kT/E_F contributes, as Figure 3.5 indicates.   Now, E_F is of the order of 5 eV for most metals, while kT at room temperature is about $\frac{1}{40}\,eV$.   Thus, at ordinary temperatures, less than 1 per cent of the valence electrons are responsible for the electronic specific heat. The electronic specific heat contribution can be approximated by assuming that each of the $N(kT/E_F)$ electrons which contributes absorbs the classical kinetic energy $\frac{3}{2}\,kT$ (Equation 3.2b)

$$c_{v\text{ electron}} \cong \left(\frac{3Nk}{E_F}\right)T \tag{3.17}$$

where N is the number of free electrons per mole. Thus, the electronic specific heat is small and increases linearly with temperature. Its effect is important at low temperatures when the lattice specific heat goes to zero, and at high temperatures in electropositive metals.

3.4   OTHER FACTORS CONTRIBUTING TO THE SPECIFIC HEAT OF SOLIDS

In addition to the lattice vibration and the electron contributions to the specific heat, other processes may contribute to energy absorption in certain solids. Such processes as the destruction of long-range atomic arrangement in alloys (order-disorder), the randomization of electron spins in a ferromagnetic material (see Chapter 9), and the change in distribution of electrons in a superconductor (see Chapter 11), all lead to an increase in specific heat. These so-called second-order transformations produce a cusp (or Lambda point) in the specific heat curve. Above the temperature at which randomization of the process is complete, the specific heat returns to its normal value.

3.5   THE TOTAL HEAT CAPACITY

The total heat capacity of an ideal crystal can be expressed as the sum of two terms:

$$c_{v\text{(total)}} = c_{v\text{(lattice)}} + c_{v\text{(electrons)}} \tag{3.18}$$

At low temperatures, Equations 3.15 and 3.17 can be used together:

$$c_v = AT^3 + \gamma'T \tag{3.19}$$

where A and γ' represent the constants of Equations 3.15 and 3.17. To determine these constants from low temperature heat capacity data, another form of Equation 3.19 is sometimes used:

$$\frac{c_v}{T} = \gamma' + AT^2 \qquad (3.20)$$

When the quantity c_v/T is plotted as a function of T^2, a straight-line plot results. The intercept and slope give γ' and A, respectively, as seen in Figure 3.6. In practice, the lattice heat capacity is usually much larger than the electronic heat capacity except at very low temperatures (about 1 to $2°K$).

We have shown above the calculation of c_v, the heat capacity, at constant volume; constant volume is the simplest boundary condition. In the laboratory, however, the heat at constant pressure is usually measured, since the maintenance of constant volume under changing temperature conditions is extremely difficult. To relate c_p to c_v, the thermodynamic relation

$$c_p - c_v = TV\frac{\alpha_v^2}{\beta} \qquad (3.21)$$

is employed. Here, V is the molar volume, and α_v and β are the experimentally determined volume coefficients of expansion and compressibility:

$$\alpha_v \equiv \frac{1}{v}\left(\frac{dv}{dT}\right)_p \qquad (3.22)$$

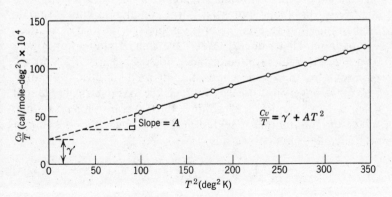

Figure 3.6   Low-temperature heat capacity of niobium.   (After F. J. Morin, and J. P. Maita, *Phys. Rev.*, Vol. 129, No. 3, 1963, p. 1115.)

and

$$\beta \equiv \frac{1}{v}\left(\frac{dv}{dp}\right)_T \qquad (3.23)$$

Table 3.2 lists the specific heat, c_p, of various elements at room temperature. If covalent primary bonds and weak secondary bonds are present (polymers and ceramics), or large magnetic moments are involved (Ni, Fe, Co), large additional contributions can dominate c_p as shown in Table 3.2.

3.6 THERMAL EXPANSION

Almost all solids expand on heating, in the absence of phase transformations. The volume coefficient of expansion α_v (Equation 3.22) is three times the linear thermal coefficient of expansion α_L,

$$\alpha_L = \frac{1}{l}\left(\frac{dl}{dT}\right)_p \qquad (3.24)$$

(see Problem 3.4).

On an atomic basis, thermal expansion corresponds to an increase in the average interatomic distance. As temperature increases, the amplitude of atomic (lattice) vibration increases, becoming as large as 12 per cent of the interatomic spacing at the melting point. This fact alone is *not* sufficient to account for thermal expansion. Consider Figure 3.7 where we have plotted a symmetric, parabolic potential corresponding to the harmonic oscillator used in Section 3.2 for the lattice specific heat derivation. Although the higher energy levels have high amplitudes, in Figure 3.7 the *average* interatomic distance does not change. Figure 3.7 represents an approximation commonly used because it is mathematically convenient. Closer to fact is the asymmetrical potential discussed in Chapter 1 of Volume I. Generally, for crystalline solids, the interatomic potential may be expressed as the sum of attractive and repulsive terms:

$$V_{\text{total}} = -\frac{A}{R^n} + \frac{B}{R^m}, \qquad (3.25)$$

where $n = 1$ (coulomb attraction) for ionic crystals. Figure 3.8 is a potential curve for an ionic crystal with m set equal to three.

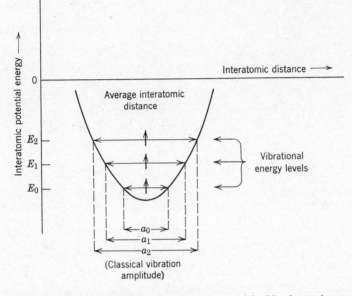

Figure 3.7   Vibration of atoms in a symmetric potential.   No change in average interatomic distance as energy increases.

Vibrational energy levels are shown schematically on the asymmetric potential curve.   It is apparent that the average interatomic distance increases as the vibrational energy increases, due to the asymmetric nature of the curve.   In covalently bonded materials, the attractive force is stronger, and n of Equation 3.25 has a higher value.   This makes the right side of the curve in Figure 3.8 steeper, restores some of the symmetry, and leads to a smaller thermal expansion.   For weaker bonds, such as the metallic bond, the right side of the curve is shallower, and a large expansion coefficient results.   For complex materials such as polymers, the above analysis is not applicable.   Table 3.2 lists the coefficients of thermal expansion for a number of materials, along with the melting points; for simply bonded materials, the melting point is a good measure of the attractive potential.   In metals where the structures are relatively simple, Table 3.2 shows a general decrease in α_L as the melting point increases.   In some polymers (elasto-

Table 3.2   Thermal Properties for Several Materials

MATERIAL	SPECIFIC HEAT c_p (AT 300° K) CAL/g°K	ATOMIC MOLECULAR WEIGHT, g/MOLE
Aluminum	0.22	27.0
Carbon, graphite	0.18	12.0
diamond	0.12	—
Copper	0.092	63.5
Gold	0.031	197.0
Iron	0.11	55.9
Lead	0.32	207.2
Molybdenum	0.065	96.0
Nickel	0.13	58.7
Niobium	0.074	92.9
Platinum	0.031	195.1
Silver	0.056	107.9
Tantalum	0.036	181.0
Tin	0.54	118.7
Tungsten	0.034	183.9
Type 304 stainless steel (austenitic)	0.14	—
Invar	0.12	—
Alumina ceramic	0.18	101.9
Boron nitride	—	24.8
Magnesia (MgO)	0.21	40.3
Fused silica (or Fused quartz)	0.19	60.1
Glass	0.2	—
Mica	0.12–0.25	—
Bakelite	~0.4	—
Lucite	0.35	—
Teflon	0.25	—
Low-density polyethylene	0.5	—
Natural rubber	—	—
Silicone rubber	—	—

LINEAR COEFFICIENT OF EXPANSION α_L (at 300° 1/°K $\times$ 10^6)	GRUNEISEN'S CONSTANT γ	WIEDEMANN-FRANZ RATES $L = \dfrac{\sigma_T}{\sigma_e T}$ (volts/°K)^2 $\times$ 10^8	MELTING POINT T_m °K
24.1	2.17	2.2	933
2.3–2.8	—	—	3760
1.2	—	—	—
17.6	1.96	2.23	1356
13.8	3.03	2.35	1336
10.8	1.60	2.47	1810
28.0	2.73	2.47	600
5.55	1.57	2.61	2880
13.3	1.88	2.2	1725
7.4	—	—	2740
8.8	2.54	2.51	2040
19.5	2.40	2.31	1333
6.7	1.75	—	3270
23.5	2.14	2.52	505
3.95	1.62	3.04	3680
17.3	—	—	~1690
1.26	—	—	~1700
6.7	—	—	2470
7.72	—	—	1950
14.0	—	—	3020
0.05	—	—	1950
7.2	—	—	—
40	—	—	—
15–45	—	—	—
81	—	—	—
100	—	—	—
180	—	—	—
670	—	—	—
1200	—	—	—

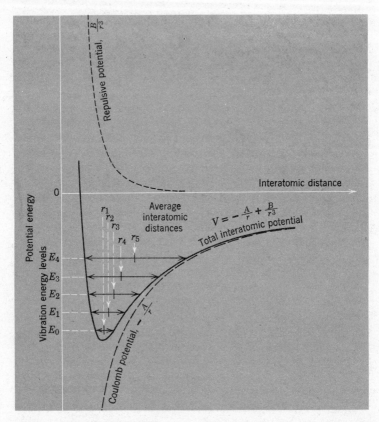

Figure 3.8   Potential energy for hypothetical ionic crystal with m-3 for the repulsion potential. For real ionic crystals, m is much larger, usually about 10. As the energy of vibration increases, the average interatomic distance increases due to the asymmetric potential.

mers), the extremely large thermal expansion found arises from the very weak secondary bonds and sparse cross-linkage between the molecules. Thermoplastic polymers, such as lucite, have smaller values of α_L, as the result of stronger cross-linkage. Bakelite, which is thermosetting, has the lowest α_L compared to the other polymers in the table. This may be attributed to the covalent cross-linkage present. Table 3.2 lists only averages found for polycrystalline materials. If we are dealing with single crystals, we should bear in mind that because thermal expansion is so intimately connected to bond strength, α_L is a function of crystallographic orientation.

It is possible, using the Debye model, to derive an expression for linear expansion which includes the molar heat capacity c_v and the compressibility β (see Equation 3.23) of a solid of molar volume V:

$$\alpha_L = \frac{\gamma c_v \beta}{3V}, \qquad (3.26)$$

where γ is Gruneisen's constant. The latter is defined by the equation

$$\gamma = -\frac{d \ln \nu}{d \ln V_a} \qquad (3.27)$$

where ν is the atomic vibration frequency and V_a is the atomic volume. Gruneisen constants are listed for various materials in Table 3.2.

The coefficient of expansion of fused quartz is normally extremely low and actually becomes negative at low temperature. As a consequence of its low coefficient of thermal expansion, fused silica (quartz) has a high thermal shock resistance. Most other glasses and ceramic materials fail by heat shock.

The low coefficient of expansion of the Fe-Ni (36 per cent Ni) alloy, Invar, occurs because its tendency to expand on heating is nullified by its contraction due to magnetostriction as it approaches the Curie point (see Chapter 9). Above its Curie point ($\sim 200°C$), it expands normally. Note that from Equation 3.21, the specific heat will also be affected by loss of magnetism. This effect was discussed in Section 3.4. Special low-expansion alloys of this type have been developed for applications where control of thermal expansion is necessary.

3.7 THERMAL CONDUCTIVITY

The rate at which heat is transferred through a solid is determined by the thermal conductivity of the solid, σ_T, and the temperature gradient, dT/dx. Figure 3.9 shows a schematic arrangement for measurement of heat conduction. If the heat flux, q, is defined as the calorie flow per unit time across a unit cross-section;

$$H/A = q = -\sigma_T \frac{dT}{dx} \qquad (3.28)$$

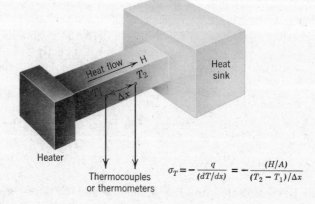

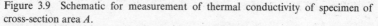

$$\sigma_T = -\frac{q}{(dT/dx)} = -\frac{(H/A)}{(T_2 - T_1)/\Delta x}$$

Figure 3.9   Schematic for measurement of thermal conductivity of specimen of cross-section area A.

The minus sign indicates that heat flows from higher to lower temperatures.   Note that the equation for heat conductivity is the same as that for electrical conductivity (Ohm's law Chapter 4), and diffusion (Fick's first law, Volume II, Chapter 5).

The thermal conductivity of a single crystal will depend on the crystallographic direction.   The values of thermal conductivity used for engineering purposes are usually averages for random polycrystalline materials.   Table 3.3 and Figure 3.10 show typical data for σ_T.   Often, in the case of rolled or drawn (wire) materials, a majority of the grains are aligned in one crystallographic direction; such materials are said to be "textured" (this is discussed further in Chapter 10 and Figure 10.4).   The single crystal anisotropy σ_T will consequently show up in textured materials as well.

For metals at room temperature, good thermal conductivity is always accompanied by good electrical conductivity because heat transfer in metals is due mainly to free electrons.   In ionic or covalently bonded solids where electrons are not as mobile (see Chapter 4), heat is transferred mainly by phonons.   Although phonons move with the speed of sound, they collide relatively often with each other and with lattice defects.   Consequently, σ_T is much lower than the phonon velocity would lead one to expect. In polymers, heat transfer occurs by molecular rotation, vibration, or translation.

Table 3.3   Thermal Conductivities at 300° K

MATERIAL	CAL/(cm SEC °K)
Al	0.53
Cu	0.94
Fe	0.18
Ag	1.00
C(Diamond)	1.5
Ge	0.14
Ni-Cr (70:30) s.s.	0.034
Cu-Zn (70:30) s.s.	0.24
Steel	0.12
NaCl	0.017
KCl	0.017
AgCl	0.0026
Glass	0.0019

Let us make a simple approximation for the conductivity due to electrons or phonons. One can visualize the electrons or phonons as a gas, the particles of which have some average velocity V_s. They are scattered by collisions with other electrons, phonons, impurities, or imperfections after moving some average distance l,

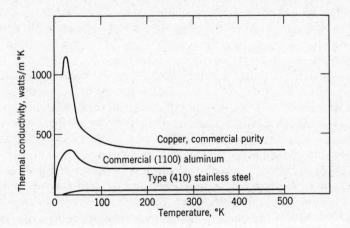

Figure 3.10   Low-temperature thermal conductivity of three commercial materials. (From the *Cryogenic Materials Data Handbook,* National Bureau of Standards, OTS PB 171809.)

giving up a portion of their energy. Thermal conductivity in solids can then be viewed as somewhat analogous to heat conduction in a gas. Following the relation for the thermal conductivity as given in the kinetic theory of gases

$$\sigma_T = \tfrac{1}{3} n c_v V_s l \qquad (3.29)$$

where n is the density of free electrons or phonons and c_v is taken here as the specific heat per electron or phonon and l is the average distance between collisions.

For good electronic heat conductors like silver or copper, both the velocity and mean path length of electrons is 10 to 100 times that for phonons. However, the electron specific heat, $n_e c_v$, is only one $1/100$ of the phonon specific heat. The net result from Equation 3.29 is that the electron conductivity is 10 to 100 times greater than the phonon conductivity in these metals.

Some very pure and perfect nonmetallic crystals which are phonon conductors show a thermal conductivity comparable to metals in certain temperature ranges. In these crystals, the phonon path is long, and there are relatively few imperfections for scattering. Thus, diamond is a better thermal conductor than silver between room temperature and $30°K$, and sapphire is a better conductor between 90 and $25°K$.

On the other hand, the thermal conductivity of large groups of metals with high solid solution alloy content is extremely low, less than one tenth that of the pure base metal. An example is the austenitic stainless steel shown in Figure 3.10; it is practically a thermal insulator at most temperatures. For such alloys, the atoms in solid solution scatter both phonons and electrons and reduce the conductivity by decreasing the mean free paths.

With increasing temperature, the average phonon and electron energies and velocities increase, while their mean free paths decrease. Except at low temperatures, the two phenomena tend to cancel one another in pure metals, and the thermal conductivity is essentially constant with temperature as shown in Figure 3.10. At low temperatures, the phonon mean free path may be limited by impurities and imperfections. In such cases, especially if the specific heat is still increasing with temperature, σ_T will increase.

In pure metals, electrons are mainly responsible for heat conduction. The Wiedeman-Franz (1853) relationship

$$\frac{\sigma_T}{\sigma_e T} = L \tag{3.30}$$

where L is approximately constant for most pure metals, expresses the relationship between electrical and thermal conductivity. Values of L are listed for a number of metals in Table 3.2. Derivation of Equation 3.30 is left as an exercise for the reader (Problem 4.23). For most metals the ratio found experimentally is somewhat larger than that given by the above equation.

The best thermal insulators are highly porous materials. This can be attributed to the low thermal conductivity of the still air in the pores. In high-temperature furnaces, porous refractories or ceramic powder insulation are used. During prolonged high temperature use most refractories tend to sinter which leads to densification and loss of insulating quality. Fine carbon powder is an exception and therefore particularly useful.

The best insulator of all is a vacuum, because heat can only be transferred by radiation. It is used to insulate liquid-helium and liquid-nitrogen containers. Porous polymeric materials approach the performance of a vacuum at low temperatures, as the gas in the pores freezes, leaving essentially evacuated voids. This type of insulation is used in various cryogenic applications.

DEFINITIONS

Heat Capacity.   The amount of heat necessary to raise the temperature of a system one degree.

Specific Heat.   The heat capacity per unit mass. Sometimes used to refer to the heat capacity per mole.

Fermi Level.   In a partially filled energy band at $0°K$, the Fermi level is the energy of the highest filled state. At higher temperatures, one half of the states at the Fermi level are full.

Electronic Specific Heat.   The contribution to the total specific heat due to transitions of electrons to states of higher energy.

Lattice Specific Heat.   The contribution to the total specific heat due to transitions of the vibrating atoms in the crystal lattice to vibrational states of higher energy.

Dulong-Petit Law.   All elemental solids approach a specific heat $c_v = 3R$ cal/mole $°K$ at sufficiently high temperatures.

Debye Temperature.   ($\theta_D = h\nu_D/k$): A parameter in Debye's formula for the lattice specific heat, proportional to the maximum phonon frequency ν_0. In practice, θ_D may be found by fitting the Debye formula to the specific heat data.

Gruneisen Constant (γ). The negative rate of change of the atomic vibration frequency with respect to atomic volume change; or

$$\gamma = -\frac{d\ln\nu}{d\ln V_a}$$

where it is assumed that γ is the same for all vibrational states within the solid.

BIBLIOGRAPHY

SUPPLEMENTARY READING:

Azaroff, L. V., and Brophy, J. J., *Electronic Processes in Materials,* McGraw-Hill, New York, 1963, Chapter 5, pp. 123–164.

Feynman, R. P., Leighton, R. B., and Sands, M., *The Feynman Lectures on Physics,* Volumes I and II, Addison-Wesley, Reading, Mass., 1964, Chapters 40–42, Volume I.

Halliday, O., and Resnick, R., *Physics,* Volume I, John Wiley & Sons, New York, 1960, Chapter 22.

Mendelssohn, K., *Cryophysics,* Interscience, New York, 1960, Chapter 5.

Weiss, R. J., *Solid State Physics for Metallurgists,* Addison-Wesley, Reading, Mass., 1964, pp. 65–85 for the theory, and pp. 268–286 for experiments.

Zwikker, C., *Physical Properties of Solid Materials,* Interscience, New York, 1954, Chapters IX and XI.

ADVANCED READING:

Mott, M. F., Jones, H., *The Theory and Properties of Metals and Alloys,* Dover, New York, 1958, pp. 1–4, 15–25, 178–183.

Smith, R. A., *Wave Mechanics of Crystalline Solids.* Chapman-Hall, London, 1961, Chapters 3, 5, 6, 7, and pp. 308–9.

Wilson, A., *The Theory of Metals,* Cambridge University Press, 1958, pp. 133–150.

PROBLEMS

3.1   (a) Discuss the reasons for the failure of the law of Dulong and Petit to predict specific heats at low temperatures.

(b) Why should the law be valid at high temperatures?

3.2   (a) If $c_p = (\partial H/\partial T)_p$ and $c_v = (\partial E/\partial T)_v$, show that $c_p - c_v = [(\partial E/\partial V) + p](\partial V/\partial T)_p$.

(b) How would you measure c_v for a solid?

3.3   Derive the average energy of a quantized harmonic oscillator $\bar{E}$, Equation 3.8.   *Hint.* $\Sigma N_n = N_0(1 + e^{h\nu/kT} + e^{2h\nu/kT} + \ldots)$ Check the sum of the series $(1 + x + x^2 + \ldots)$ in any handbook of functions. (The distribution of atoms among the various energy states is known as the "Partition Function" in statistical mechanics.)

$$\Sigma N_n E_n = N_0(h\nu e^{h\nu/kT} + 2h\nu e^{2h\nu/kT} + \ldots)$$

Check the sum of the series $(1 + x + 2x^2 + 3x^3 + \ldots)$.

3.4   Show that the Einstein specific heat (Equation 3.10) converges to $3R$ at high temperatures. (*Hint.* Starting with Equation 3.9, for $kT \gg h\nu$, the series summation of $e^{h\nu/kT} \cong 1 + h\nu/kT$, the sum of the first two terms only.)

3.5   For most of the pure elements of Table 3.2, the specific heat at $300°K$, calories per mole $°K$, is close to the Dulong-Petit prediction, while the value for diamond is much lower. Would you expect a Debye temperature above or below room temperature for these cases, and why?

3.6   The measured specific heat of aluminum at $25°K$ is 0.11 calories per mole $°K$. Given the Einstein and Debye characteristic frequencies of 6.42×10^{12} and 8.26×10^{12} per second, respectively, for aluminum, calculate the Einstein and Debye specific heats at $25°K$.

3.7   From Figure 3.6, determine the electronic and lattice specific heats for niobium at $16°K$.

3.8   Describe, with the aid of a graph, the effect of the order-disorder transformation in β brass (CuZn) upon the specific heat. (See Mendelssohn, p. 40.)

3.9   Indicate, graphically, the vibrational energy spectrum of a solid according to (a) Einstein (1 frequency), (b) Nernst-Lindemann (two frequencies), (c) Debye and (d) Blackman (see Mendelssohn, *Cryophysics*, p. 35).

3.10   Show, to a first approximation, that the coefficient of volume expansion equals three times the coefficient of linear expansion.

3.11   From Figure 3.8 and Equation 3.29, explain the physical basis for the fact that the coefficient of linear expansion of a solid approaches zero at low temperatures.

3.12   The coefficient of expansion at room temperature of crystalline quartz parallel to the hexagonal axis is 7.97×10^6, while that for fused quartz is 0.50×10^6. What is the cause of this large difference in expansion coefficient?

3.13   Explain why 18:8 stainless steel frying pans are plated with a heavy copper bottom, while aluminum frying pans are not.

3.14   The shock parameter P is a measure of the sudden temperature change which a material can withstand without cracking. It may be defined as

$$P = \frac{\sigma_T(TS)}{E\alpha_L}$$

where σ_T is the thermal conductivity, TS is tensile strength, E is Young's Modulus, and α_L is coefficient of expansion. How would you rate the shock resistance of (1) ceramics, (2) alloys, (3) cermets (ceramic-metal mixtures), and (4) pure metals.

3.15   Explain why metals will feel colder to the touch than ceramics or plastics, although all three materials are at room temperature.

3.16   Explain how the thermal conductivity may (a) increase, or (b) decrease with temperature. Which of these cases would apply to metals, alloys, and covalent and ionic solids?

3.17   Why does a porous material provide better insulation than a solid material? (See Zwikker, *"Physical Properties of Solid Materials,"* Interscience, New York, 1954, p. 165.)

3.18   The thermal conductivity of LiF made from the isotope Li[6] is similar to LiF made from the isotope Li[7]. In LiF made of 50 per cent Li[6] and 50 per cent Li[7] the thermal conductivity is only half that of the prime material. Explain (see Mendelssohn, pp. 82–84).

CHAPTER FOUR

Electrical Conduction

When an electric field is applied to a solid, free electrons are accelerated. They lose or decrease their kinetic energy by collisions with the atoms of the lattice. The current that results is proportional to the average electron velocity which is determined by the intensity of the applied electric field and the collision frequency. Only electrons whose energy is near the Fermi surface may be accelerated, as the other electrons inhabit states which are surrounded by full states, and are thus forbidden to accelerate (change state) by the Pauli principle. If the valence band is full and does not overlap empty bands, the lack of adjacent empty states severely limits conduction as in insulators and semiconductors. The wave-mechanical model for the electron in a crystal shows that the response to the applied field can be determined from the E versus κ relation; in certain cases, the electron may act as if it had a negative mass. Electrons may move through an ideal crystal without resistance, but in real crystals electrons collide with phonons, dislocations, vacancies, impurity atoms, and any other lattice imperfection. The resistivity due to solute atoms, impurities and imperfections is called the residual resistivity and is usually independent of temperature. The residual resistivity due to any one impurity or imperfection is, for dilute concentrations, proportional to the concentration. The total resistivity is the sum of the residual and thermal contributions. In ionic solids where the electrons do not participate in electrical conduction, charge transport occurs by ion diffusion.

4.1 INTRODUCTION: A SIMPLE CLASSICAL MODEL

Electrical currents flow in a conducting solid when an electrical potential differences exists. Usually the current density J_e is proportional to the electrical field:

$$\mathbf{J}_e = \frac{\mathscr{E}}{\rho_e} = \sigma_e \mathscr{E} \qquad (4.1)$$

In Equation 4.1, ρ_e is called the electrical resistivity and its reciprocal σ_e the electrical conductivity. The current density J_e is the rate of passage of charge (coulombs per second) per unit area (meters)^2 in MKS units. The field $\mathscr{E}$ is measured in volts/meter. Equation 4.1 is another way of expressing Ohm's law. The schematic of Figure 4.1, shows a way to measure electrical conductivity, analogous to the thermal conductivity experiment described in Chapter 3. The current I passes through the rod of constant cross-section A and the voltage difference $(V_2 - V_1)$ is measured between the two points. The current density J_e in the rod is I/A, the electrical field is $(V_2 - V_1)/\Delta x$, and

$$J_e = \frac{I}{A} = \frac{\sigma_e(V_2 - V_1)}{\Delta x} \qquad (4.2)$$

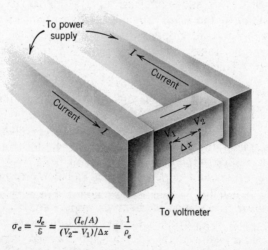

$$\sigma_e = \frac{J_e}{\mathscr{E}} = \frac{(I_e/A)}{(V_2 - V_1)/\Delta x} = \frac{1}{\rho_e}$$

Figure. 4.1   Measurement of electrical conductivity of a specimen of cross-sectional area A.

Ohm's law is usually written as

$$V = IR \qquad (4.3)$$

which gives the voltage drop across an electrical resistance R when the total current I passes. According to Equation 4.2 the resistance of a length Δx of the bar in Figure 4.1 is

$$R(\Delta x) = \frac{\Delta x}{\sigma_e A} = \frac{\rho_e \Delta x}{A} \qquad (4.4)$$

The electrical resistance is a function of geometry, but the resistivity ρ_e is a material constant. Resistance is measured in units of ohms; resistivity, in ohm-meters; and conductivity, in mhos per meter.

Pure metals, as shown in Table 4.1, are the best electrical con-

Table 4.1   *Typical Electrical and Thermal Conductivities at Room Temperature*

MATERIAL	THERMAL CONDUCTIVITY σ_T OR K (WATTS/m-°C)	ELECTRICAL CONDUCTIVITY σe (mho/m)
Silver, commercial purity	420	6.3×10^7
Copper, OFHC	390	5.85×10^7
Copper + 2% Be	180	2.0×10^7
Gold	290	4.25×10^7
Aluminum, commercial high purity	230	3.5×10^7
Aluminum + 1% Mn	192	2.31×10^7
Brass, yellow	115	1.56×10^7
Tungsten, commercial	167	1.82×10^7
Ingot, iron, commercial	66	1.07×10^7
1010 Steel	47	0.7×10^7
Nickel, commercial	62	1.03×10^7
Stainless steel, type-301	16	0.14×10^7
Graphite	170 (av)	10^5 (av)
Window glass	0.9	2–3×10^{-5}
Bakelite	0.43 (av)	1–2×10^{-11}
Lucite	0.17	$10^{-17} - 10^{-14}$
Borosilicate glass	1.15	$10^{-10} - 10^{-15}$
Mica	0.51	$10^{-11} - 10^{-15}$
Polyethylene	0.33	$10^{-15} - 10^{-17}$

ductors. In metals the flow of charge is due to the motion of free electrons which are accelerated by an applied electrical field. Table 4.1 also lists thermal conductivities σ_T which are again quite large for metals because electrons carry heat readily as well as charge. At the bottom of Table 4.1 are the insulators, i.e., materials of measureable yet low conductivity. That Table 4.1 covers a range of 10^{25} in electrical conductivity, deserves comment. For some reason, all of the electrons in the insulators are not free to move.

Consider an electron which can move freely in a solid. An applied field $\mathcal{E}$ results in a force $e\mathcal{E}$ which accelerates the electron according to

$$a = \frac{e\mathcal{E}}{m_e} \tag{4.5}$$

since the electron does not accelerate indefinitely (the physical consequences would be remarkable), we can suppose that it regularly loses all its kinetic energy by collisions with phonons, impurities, or imperfections. If the average time between collisions is 2τ (τ is called the *relaxation time*), the average velocity of the electron is

$$\bar{v} = \frac{\tau e\mathcal{E}}{m_e} \tag{4.6}$$

Figure 4.2 indicates the velocity of the free electrons as a function of time according to this model. If the density of free electrons is n, the current density is

$$J_e = ne\bar{v} = \frac{ne^2\mathcal{E}\tau}{m_e} \tag{4.7}$$

and, substituting Equation 4.1 gives

$$\sigma_e = \frac{ne^2\tau}{m_e} \tag{4.8}$$

Even with no applied field the conduction electrons move about quite rapidly. The thermal energy changes discussed in Section 3.3 are essentially changes in kinetic energy; however, the electrons move about at random and there is no *net* flow of electrons. An applied electric field accelerates all the electrons in the same direction and does give a net electron flow as well as a gradual increase in kinetic energy. However, for an electron far below the Fermi

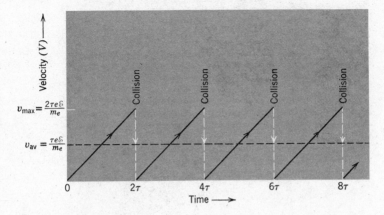

Figure 4.2   Velocity versus time for classical model of electrical conductivity of a free electron in a solid.

level, however, all the adjacent states are filled, and its energy can increase only if a sufficiently large quantity of energy is supplied to boost it above the Fermi level. A gradual acceleration cannot occur unless the electron is above or just below the Fermi level. As stated, without proof, in Chapter 3, if the electron has the energy $E_F - kT$ or above, it can usually find empty states. Thus, the density n of "free" electrons at room temperature is about 1 per cent of the valence electron density. Also, as pointed out in section 4.3 the response of an electron in a solid to an applied electric field is not as simple as that of a free electron.

4.2   BAND MODEL OF CONDUCTIVITY

On bringing atoms together to form a solid, as shown in Chapter 1, the outer or valence levels are the ones more likely to be broadened. The inner levels, the atomic or core levels, are not. As mentioned before (Chapter 1), each valence level in a solid of N atoms splits into N levels. Thus, an s level, which can contain two electrons of opposite spin, becomes an s band with room for $2N$ electrons. A p level becomes a p band with room for $6N$ electrons. The uppermost band, if it is partially filled and contains the Fermi level, is called the conduction band, because only the

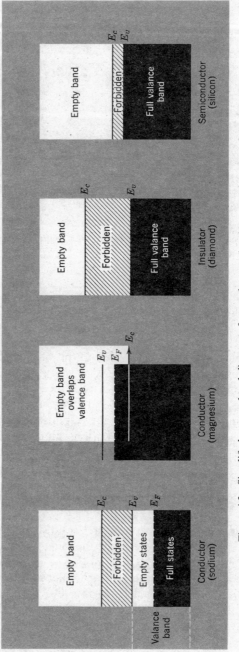

Figure 4.3 Simplified energy band diagrams for conductors, insulators, and semiconductors.

electrons in the vicinity of the Fermi level are free to carry electrical current.   In metals, the valence band is usually only partially filled.   In copper, for example, the 4s valence electrons fill only one half of the outer s-band, since there is only one 4s electron per atom.   In some metals, the valence band is full but overlaps a higher empty band.   Electrons near the Fermi levels of such metals are still free to move as the extra band supplies empty states.

Suppose, now, that the valence band is completely filled and the next higher band is completely empty, and there is no overlapping of the bands.   None of the electrons may respond to an applied electric field unless they have access to the empty states in the higher band.   To do this, they must cross the energy gap between the bands.   If the gap is a few electron volts wide, extremely high electric fields would be required to bring electrons to the higher band.*   In this situation, without free electrons, we probably have an insulator.   If the energy gap is small enough, the effect of thermal energy at higher temperature on the Fermi-Dirac distribution (see Chapter 3) may send a few electrons into the empty higher band and also create a few vacant states in the valence band.   Such material is a *semiconductor*.   Simplified band schemes for the three cases described above are shown in Figure 4.3.

In covalent solids, all the valence states are saturated and the valence band is full.   Pure diamond, which shows the strongest covalent bonding, is an insulator, with a $7 \, eV$ energy gap.   Silicon, which has a $1.1 \, eV$ energy gap, is a semiconductor.   The interesting and useful electrical properties of semiconductors and insulators are discussed in Chapters 5, 6, 7, and 11.

4.3   THE WAVE-MECHANICAL MODEL

Introducing the band model (Section 4.2), brings quantum mechanics into the picture of conductivity.   It helps to explain the

---

* Suppose the "mean free path" l between collisions were 500 angstrom units.   We can use $l \cong 2\tau v$, and with Equation 4.6 we can show that fields of about 1.6×10^8 volts/meter are necessary to increase the average kinetic energy by two electron volts (see Problem 4.23).   Similar considerations apply to the case of electrons far below the Fermi level and surrounded by full states.   Quantum mechanical calculations show that even higher fields are needed than our simple classical model predicts.

low conductivity of insulators and semiconductors. This approach can be extended by applying quantum mechanics to the motion of an electron in a crystalline solid. It is then necessary to consider the effect of the wave properties of the electron when an electrical field is applied. In wave mechanics the electron is regarded as a *packet of waves* rather than a particle. The velocity v_g of the packet in optics called the *group velocity*

$$v_g = 2\pi \frac{dv}{d\kappa} \tag{4.9}$$

where v is the DeBroglie frequency and κ the wave number. Since $E = hv$,

$$v_g = \frac{2\pi}{h} \frac{dE}{d\kappa} \tag{4.10}$$

If the electron is accelerated

$$a = \frac{dv_g}{dt} = \frac{2\pi}{h} \frac{d}{dt} \left(\frac{dE}{d\kappa} \right) = \frac{2\pi}{h} \frac{d^2E}{d\kappa^2} \frac{d\kappa}{dt} \tag{4.11}$$

Suppose the electron is accelerated by the applied field $\mathcal{E}$. In the time dt, the energy increase dE of the electron is (using Equation 4.10):

$$dE = \left(\frac{dE}{d\kappa} \right) d\kappa = e\mathcal{E}\, dx = e\mathcal{E}(v_g\, dt) = \frac{2\pi e\mathcal{E}}{h} \frac{dE}{d\kappa}\, dt \tag{4.12}$$

and, solving for $d\kappa/dt$ gives

$$\frac{d\kappa}{dt} = \frac{2\pi}{h} e\mathcal{E} \tag{4.13}$$

Substituting Equation 4.13 into Equation 4.11 gives.

$$a = e\mathcal{E} \frac{4\pi^2}{h^2} \frac{d^2E}{d\kappa^2} \tag{4.14}$$

For a free electron, we can use the parabolic E versus κ relation (see Section 1.1, Chapter 1, and Figure 1.2), and obtain Equation 4.5, which describes the classical behavior of an electron. This follows from Equation 1.4, since $d^2E/d\kappa^2 = h^2/4\pi^2 m_e$. We have seen in Chapter 1, however, that the periodic crystal lattice interrupts the parabolic relation with results like those shown in Figures

1.7 and 1.8. Near the forbidden band, the curvature of the E versus κ curve changes, and can become negative. The consequence, as we can see from Equation 4.14 of a negative curvature, $d^2E/d\kappa^2$ is that the electron in a crystal lattice may be accelerated in *the same* direction as the applied field. This startling behavior is really due to the diffraction effects which dominate the behavior of the electron as the forbidden energy is approached.

Comparing Equation 4.14 with the classical expression for electron acceleration (Equation 4.5), we can define the *effective mass* m^* of an electron in a lattice as

$$m^* = \frac{h^2}{4\pi^2}\left(\frac{d^2E}{d\kappa^2}\right)^{-1} \tag{4.1}$$

Then Equation 4.14 becomes $a = e\mathcal{E}/m^*$, where $m^* = m_e$ for a free electron. For the electron in a crystal, the value of m^* depends on its energy, as shown in Figure 4.4. The effective mass may be positive or negative. It is equal to m_e only when the

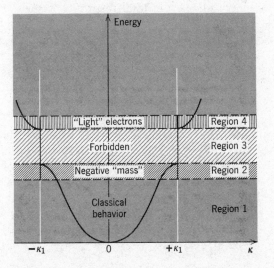

Figure 4.4 Effective mass: The use of Eq. 4.15 in the E versus κ relation. In region *1*, the parabolic relation is good, and behavior is therefore classical. In region *2*, the curvature is negative, and therefore the effective mass is negative. In region *4*, the curvature is positive but larger than that of region *1*; therefore, the effective mass is smaller, according to Eq. 4.15.

energy is not near the edge of a band and the E versus κ curve is parabolic. In most conductors, $m^* = m_e$ as the band is only partially filled. In semiconductors, insulators, and certain conductors (e.g., bismuth) where full or almost full valence bands are involved, the effective mass differs from m_e.

4.4   THE ELECTRICAL RESISTIVITY OF CONDUCTORS

According to wave mechanics, a free electron can move in a perfect crystal lattice without loss of energy. Any disturbance of the crystal structure, any perturbation of the atoms from their lattice sites, will scatter electrons and thereby cause electrical resistance. Phonons cause such a disturbance. Even at $0°K$, electrical resistance is observed in a real conductor since impurities, grain boundaries, dislocations, vacancies, and other imperfections abound in all real materials. All of these will scatter electrons. In pure metals and dilute alloys, the total resistivity is the sum of two terms: the thermal component ρ_T, which arises from the lattice vibrations, and the *residual resistivity* ρ_r, caused by impurities and structural imperfections. The latter is independent of temperature. The total resistivity is given by

$$\rho = \rho_T + \rho_r = \frac{1}{\sigma_e} \tag{4.16}$$

The Equation 4.16 is *Matthiessen's rule*. Experimental verification of the rule is shown in Figure 4.5 for a series of dilute Cu-Ni alloys, the residual restivity rising as the nickel content of the alloy increases. Matthiessen's rule becomes less accurate at high temperatures or at high impurity levels.

Above the Debye temperature, the thermal component of resistivity of conductors is approximately linear. The resistivity can therefore be expressed as

$$\rho = \rho_0(1 + \alpha(T - T_{R.T}) + \cdots) \tag{4.17}$$

where ρ_0 is the room temperature resistivity ($1/\sigma_e$ in Table 4.1). For pure metals, the temperature coefficient of resistivity, α, is about 0.004 per degree centigrade, while for alloys, α is generally

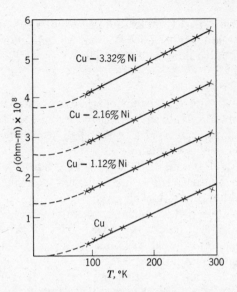

Figure 4.5   Electrical resistivity of Cu-Ni alloys as a function of temperature. [After J. O. Linde, *Annalen der Physik*, Vol. 5, p. 15 (1932).]

lower.   At very high temperatures, Equation 4.17 does not apply, since other scattering processes occur.

A simple way to estimate the over-all purity and perfection of a conductor is to measure the ratio of the resistivities at room temperature and at liquid helium temperature, i.e., $\rho(298°\text{K})/\rho(4.2°\text{K})$.   At $4.2°\text{K}$, $\rho \cong \rho_r$ and therefore the resistivity ratio is approximately $(\rho_T(298) + \rho_r)/\rho_r$.   For very pure and structurally perfect metals, the resistivity ratio can be very large. Values of 100,000 have been achieved after zone refining (Chapter 7) and long time outgassing in vacuum at elevated temperatures. For commercial purity materials, ratios below 100 are more common; for some alloys the ratios are as low as 1.

As shown in Figure 4.5, the addition of an impurity raises the residual resistivity.   The dependence of ρ_r on a single impurity is

$$\rho_r(x) = Ax(1 - x) \qquad (4.18)$$

where x is the concentration and A is a constant which depends on the base metal and the impurity.   The value of A increases with valence, atomic size, or other differences between the two materials. Equation 4.18 is called *Nordheim's rule*. For dilute solutions, $x \ll 1$, and Equation 4.18 becomes

$$\rho_r(x) \cong Ax \tag{4.19}$$

Figure 4.6 illustrates such behavior when ρ_r is measured directly. Even at room temperature, the resistivity increases linearly with concentration, as Figure 4.7 shows. The linear increase depends on ρ_r for ρ_T remains constant.

Since residual resistivity results from electron scattering by impurities and imperfections, it is affected by changes in distribution of solute atoms, as well as concentration. It can also be influenced by cold work, quenching, or neutron irradiation. These processes introduce an excess of vacancies, dislocations, or interstitial atoms into the lattice. Heat treatments which lower or change the distribution of imperfections can lower the resistivity. Resistivity experiments are often employed to determine the rate and amount of structural change which result from the thermal and mechanical processing of metal.

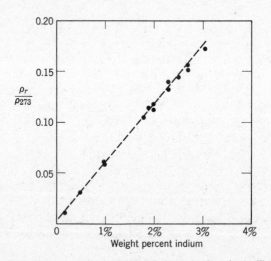

Figure 4.6   The residual resistivity of dilute solutions of indium in tin, illustrating the Nordheim rule. [After A. B. Pippard, *Proc. Roy. Soc.* (*London*), Series A, Vol. 248, p. 97 (1955).]

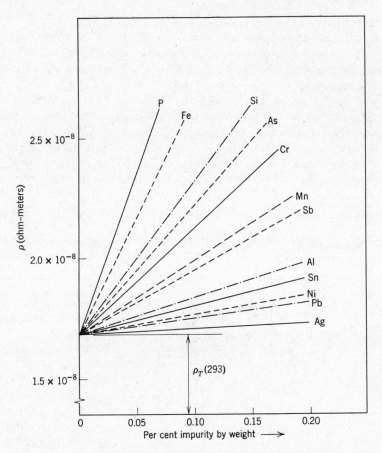

Figure 4.7   The effect of small additions of various elements on the room-temperature electrical resistivity of copper. [After F. Pawlek and K. Reichel, *Zeitschrift für Metallkunde*, Vol. 47, p. 347 (1956).]

4.5   THE ELECTRICAL RESISTIVITY OF MULTIPHASE SOLIDS

Let us now examine the dependence of electrical resistivity on composition when one or more phases may be present. In a binary alloy system with only one solid solution (complete solid solubility), we can use Equation 4.18 to build up parabolic curves from each pure component, with a maximum at 50 atomic per cent. An example of resistivity changes across a binary solid solution diagram is shown in Figure 4.8. All solid solution alloys do not necessarily exhibit a maximum near the 50 per cent composition.

In two-phase alloys, the situation is more complicated. To simplify the geometric problem, let us take as a model a random two-phase $(\alpha + \beta)$ mixture. Let V_α be the volume fraction of α present, and V_β the volume fraction of β. The resistance of a rod of the two-phase alloy, of cross-section area A and length l, is

$$R_{(\text{rod})} = \frac{\rho_e l}{A} = \frac{V}{I} \tag{4.20}$$

where V and I are the voltage across the rod and current in the rod, respectively. The rod may be divided into a bundle of N parallel

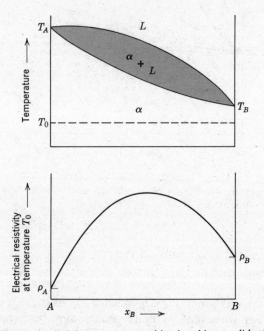

Figure 4.8   Electrical resistivity versus composition in a binary solid solution alloy system.

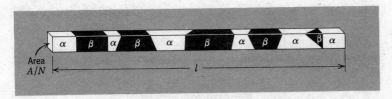

Figure 4.9   Fiber of infinitesimal cross-sectional area, sectioned from a two phase-resistivity specimen.

fibers of length l and small cross section; consider N to be a large number, and the cross section of each fiber A/N as shown in Figure 4.9. Very few of the fibers have phase boundaries parallel to the rod axis. The fiber can then have a certain volume of α in *series* with a certain volume of β. The length of β fiber present is $V_\beta l$, and α fiber, $V_\alpha l$. The resistance of the two in series is given by

$$R_{(fiber)} = \frac{\rho_\alpha V_\alpha l}{(A/N)} + \frac{\rho_\beta V_\beta l}{(A/N)} \tag{4.21}$$

Since the total rod consists of N small fibers in parallel,

$$\frac{1}{R_1} + \frac{1}{R_2} + \cdots + \frac{1}{R_N} = \frac{1}{R_{(rod)}} = \frac{N}{R_{(fiber)}} = \frac{A}{(\rho_\alpha V_\alpha + \rho_\beta V_\beta)l} \tag{4.22}$$

Combining Equations 4.20 and 4.22 gives

$$\rho_e = \rho_\alpha V_\alpha + \rho_\beta V_\beta \tag{4.23}$$

Thus, the electrical resistivity of a two-phase material is a linear function of the volume fractions of the two phases. If the densities are not too different, mass fractions can be used instead of volume fractions in Equation 4.23. The resistivity as a function of composition can then be computed across an eutectic alloy system, as shown schematically in Figure 4.10. If an intermediate phase is present as shown in Figure 4.11, it may be treated as another solid solution. The mixture law of Equation 4.23 always predicts a lower value of ρ_r than the actual value for the intermediate phase. In Figures 4.10 and 4.11, the parabolas are steeper than in Figure 4.8 because limited solid solubility implies (see Section 7.4 of Volume I or Section 2.4 of Volume II) differences in atomic size,

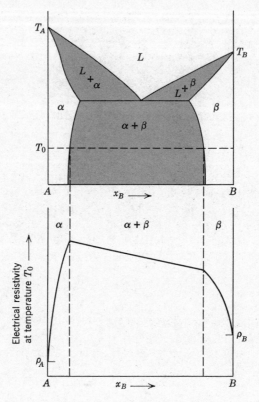

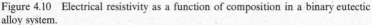

Figure 4.10   Electrical resistivity as a function of composition in a binary eutectic alloy system.

valence, crystal structure or electronegativity.   All of these factors tend to make dissolved atoms more effective scatterers of conduction electrons, thereby raising the resistivity.

4.6   THE ELECTRICAL RESISTIVITY OF IONIC SOLIDS

Although the band model also applies to ionic solids, the number of electrons occupying states in their conduction band is small and if current flows, it is more likely due to the motion of ions

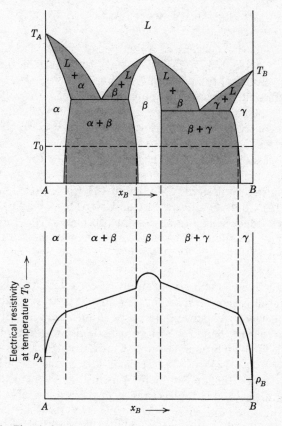

Figure 4.11   Electrical resistivity for a binary system with an intermediate phase.

by diffusion, rather than electrons.   The diffusion of ions is impossible unless interstitial ions or vacant ionic sites are present. Ionic conductivity may be calculated from the diffusion coefficient for the ion in question by using an equation for ionic conductivity of the form:

$$\sigma_i = Ne^2 \frac{D}{kT} = \frac{Ne^2}{kT} D_0 e^{-Q/kT} \qquad (4.24)$$

where Q is the activation energy for diffusion, and N is the density of ionic sites (of one sign) per unit volume.

4.7   ELECTRICAL RESISTOR MATERIALS

The resistive properties of a number of metals and alloys are practically applied in the manufacture of resistance thermometers, precision electrical resistors, and heating elements. The temperature coefficient of resistivity (Equation 4.17) of pure annealed Pt is used to define the high-temperature scale between the melting points of pure elements. Cu and Ni are also used for resistance thermometers.

For precision resistors in electronic devices, a very small temperature coefficient of resistivity in the vicinity of room temperature is normally desired. Manganin (83 Cu-13 Mn-4 Ni), constantan (55 Cu-45 Ni), or a 72 Fe-23 Cr-5Al-0.5 Co alloy are used in a stress-relieved condition for this purpose. These alloys are metallurgically stable, corrosion resistant, and show a temperature coefficient $\alpha < \pm 20 \times 10^{-8}$ per °C between 0° and 100°C.

For heating-element applications, the oxidation resistance of the material is more important than its electrical resistance. Although it is very expensive, Pt is sometimes used as a high-temperature furnace winding. More commonly, resistance elements are made from nichrome (80 Ni-20 Cr) or, for higher temperatures, 55 Fe-37 Cr-8 Al alloy. The semiconductors silicon carbide and molybdenum disilicide can be used as heating elements at higher temperatures. Oxidation resistance in these materials is insured by the silicon dioxide film which forms on the surface at elevated temperatures. To achieve even higher temperatures by resistive heating requires the use of W, graphite, or Ta elements. W and graphite may be employed in reducing atmospheres, but Ta only in a vacuum.

DEFINITIONS

Electric Current (I). The time rate of passage of charge through a conductor. In mks units, coulombs per second, called amperes.

Current Density (J). The electric current per unit area, amperes per square meter.

Ohm's Law. The current density is proportional to the electric field:
$$J_e = \sigma_e \mathscr{E}$$

Electrical Conductivity (σ_e). The proportionality constant in Ohm's Law as stated above.

Electrical Resistivity (ρ_e). The inverse of the electrical conductivity, σ_e, so that Ohm's Law is also $\mathscr{E} = \rho_e J_e$

Ohm. The unit of electrical resistance, where $V = IR$ for the entire conductor. The units are volts/ampere. Thus, resistivity is in units of ohm-meters.

Mho. The unit of electrical conductance, the inverse of the ohm. Units for the mho are ampere/volt. The conductivity σ_e is then in units of mhos/meter.

Insulator. A material whose σ_e is very low at room temperature, usually less than 10^{-6} mho/m.

Semiconductor. σ_e for a semiconductor may lie somewhere between the values for conductors and insulators; it is highly temperature and field dependent.

Relaxation Time (τ). About one half the average time between collisions of the conduction electrons with phonons, lattice defects, etc.

Conduction Band. The lowest energy band which is not completely filled with electrons. Electrons in the conduction band of a conductor are generally free to move if they are close enough to the Fermi level.

Valence Band. The band containing the valence electrons. In a conductor, the valence band is also the conduction band.

Scattering. Change of direction and energy of motion due to a collision; in quantum-mechanical terms, a transition to a new state due to a collision.

Effective Mass (m^*). The effective mass of a conduction electron is determined by its response to an applied field, $m^* = e\mathscr{E}/a$. Due to quantum mechanical effects, m^* may be large, small, positive, or negative.

Residual Resistivity (ρ_r). The temperature-independent part of the resistivity of a conductor. Lattice defects and impurities are responsible for ρ_r.

Resistivity Ratio. Usually, $\rho(298°K)/\rho(4.2°K)$, although the lower temperature may vary; $(\rho_T(298°) + \rho_r)/\rho_r$ is approximately $\rho_T(298°)/\rho_r$ in order to estimate ρ_r.

Matthiessen's Rule. The total resistivity of a conductor is the sum of a temperature-dependent component ρ_T and a temperature-independent component ρ_r.

Nordheim's Rule. $\rho_r(x) = A(x)(1 - x)$ where x is the concentration of a given impurity, and $\rho_r(x)$ is the contribution due to the impurity. A is a constant, and depends on the impurity and the conductor.

BIBLIOGRAPHY

SUPPLEMENTARY READING:

Azaroff, L. V., and Brophy, J. J., *Electronic Processes in Materials,* McGraw-Hill, New York, 1963, pp. 146–189.

Cottrell, A. H., *Theoretical Structural Metallurgy,* St. Martin's, New York.

Stanley, J. K., *Electrical and Magnetic Properties of Metals,* American Society for Metals, 1963.   Chapters 1, 2, and 4.

Wert, C. A., and Thomson, R. M., *The Physics of Metals,* McGraw-Hill, New York, 1964, pp. 205–214, 217–222.

ADVANCED READING:

Kittel, C., *Introduction to Solid State Physics,* Wiley, New York, 1953, Chapters 6, 10, and 11.

Mott, N. F., and Jones, H., *The Theory of the Properties of Metals and Alloys,* Dover, New York 1958, Chapters I, II, and VII.

PROBLEMS

4.1   A wire 0.80 in. in diameter must carry 100 amp.   However, the maximum allowable power dissipation is 10 watts per meter of wire. (Power $= IV = I^2R$).   Of the materials in Table 4.1, which are suitable for this wire?

4.2   To carry 10 amp with a maximum voltage drop of 0.5 volts per meter, what must the diameter of (a) silver, (b) copper, (c) yellow brass, (d) type 301 stainless steel wires be?   Use Table 4.1.

4.3   Suppose you needed a thermal insulator of high electrical conductivity.   What material would you specify?   Could you do the job?

4.4   A copper wire of cross-section area 5×10^{-6} square meters carries a steady current of 50 amp.   Assuming one free electron per atom, calculate: (a) the density of free electrons, (b) the average drift velocity, and (c) the relaxation time τ.

(Density of Cu $= 8.9 \times 10^3$ kg/m^3; atomic weight of Cu $= 64$.)   Use Table 4.1 if necessary.

4.5   How could you increase or decrease the conductivities of the materials in Table 4.1 on a practical basis?

4.6   In a linear polymer, how would you expect cross-linking and the degree of polymerization to affect σ_e and σ_T?

4.7   How would you expect vitrification of quartz, i.e., glass-forming, to affect σ_e and σ_T?

4.8   In a conductor with a single s electron per atom in the conduction band, m^* is approximately equal to the classical mass m.   Why?   See

Figure 4.4.   Under what circumstances of occupation of the valence band would m^* be negative?   In what type of material would you expect such effects?

4.9   What variables determine the relaxation time τ?   How?

4.10   Why does σ_e for an insulator *decrease* with increasing impurity content and temperature?   Why do conductors behave in the opposite manner?

4.11   Consider Figure 6.13 of Volume I, which refers to changes in the properties of copper on plastic deformation, and subsequent annealing. Explain the variation of the electrical resistivity.

4.12   How would you increase the resistivity ratio of a conductor?   What practical measures could you take?

4.13   The residual resistivity of a piece of highly purified Cu is increased from 10^9 mho/m to 10^7 mho/m by addition of one part per million of impurity, of a given kind.   Assuming $\rho = \rho_T + \rho_r$ at room temperature, and using Table 4.1 for the room temperature resistivity, how has the resistivity ratio changed?   How will it change if 20 parts per million are added?   How good was the assumption?

4.14   Refer to Problem 4.13.   If the purified Cu is plastically deformed 50 per cent, the residual resistivity increases to 10^6 mho/m.   If 20 parts per million of the impurity are first added, and then 50 per cent deformation done, calculate and the resistivity ratio.

4.15   (a) How does the electrical resistivity of FCC single crystal metals change with crystallographic direction?

(b) How does the electrical resistivity of hexagonal single crystal metals change with crystallographic direction?

(c) How do these values compare with that for random textured poly-crystalline metal of the same kind?

(*Reference.* Stanley, *Electrical and Magnetic Properties,* ASM, 1963, p. 51.)

4.16   Show with the aid of simple graphs how the electrical resistivity of a 4.5 per cent Cu containing aluminum alloy would change after each processing step listed below.

(a) As sand-cast.
(b) As hot-forged.
(c) As solutionized, quenched and measured at $0°C$.
(d) After cold-working at $0°C$.
(e) After holding at R.T. more than 24 hours.

4.17   Some metals show a reduction of electrical resistivity with pressure and some show an increase (Li and Bi).   How can this be explained according to band theory?

4.18   Quenching aluminum wires from 504°C into cold water results in an increase of 0.18 per cent in resistivity if measured immediately. If allowed to stand, the resistivity increase disappears in 25 minutes. Explain.

4.19   If the aluminum used in the above problem contains 0.13 per cent Mg, the resistivity does not change after quenching and holding at R.T. but does after heating at 120°C. Explain.

4.20   Thin metal films evaporated onto a cold substrate in vacuum have a resistivity about twice that of bulk density metal. This resistivity is lowered if the films are allowed to age. Explain.

4.21   We have tacitly assumed that the conduction band in an ionic or covalent solid is empty. Why?

4.22   (a) Derive the Wiedemann-Franz ratio, $\sigma_T/\sigma_e T = L$, (Equation 3.30) from Equations 3.29 and 4.8. Assume the classical values for the specific heat per electron, $c_v = \frac{3}{2}$ and for the mean electron kinetic energy, $\frac{1}{2}mv^2 = \frac{3}{2}RT$. The mean path length $l = 2\tau v$.

(b) Compare the derived value of L with measured values in Table 3.2. *Note.* This simplified derivation of the Wiedemann-Franz ratio was one of the first triumphs of the free electron theory of conductivity.

4.23   Derive the relation between the applied electric field and the increase in electronic kinetic energy, mentioned in the footnote to Section 4.2. Use classical expressions for all quantities. How accurate would you expect such a calculation to be?

CHAPTER FIVE

Semiconductors

Semiconducting materials are comparatively poor conductors of electricity. They are characterized by a nearly full valence band, which is separated from a nearly empty conduction band by an energy gap of 2 eV or less. Thermal energy can promote electrons to the conduction band and leave vacant states, called *holes*, in the valence band. Both conduction electrons and holes serve as charge carriers. When only valence and conduction states are involved, the semiconductor is said to be intrinsic. A substitutional impurity atoms of higher or lower valence may contribute electrons to the conduction band or holes to the valence band. If they make a dominant contribution to the conductivity, the material is said to be *n-type* or *p-type extrinsic,* respectively. The holes act in many respects as carriers of positive charge. The behavior of holes can be explained by using the concept of negative effective mass for the valence electrons. The *mobility* of a charge carrier is limited by collisions with impurity atoms and phonons. Excess carriers can be created by radiation and annihilated at *recombination centers*.

5.1 INTRODUCTION

Let us first reconsider the band model for electrons in an insulator such as that shown in Figure 5.1. The core levels and valence bands are full and the conduction band is empty. The passage of current requires transition of electrons to states of higher energy. Such transitions can occur only if the energy gap between the conduction and valence bands is surmounted. If the gap is a few electron volts or more, very high fields will be required to surmount

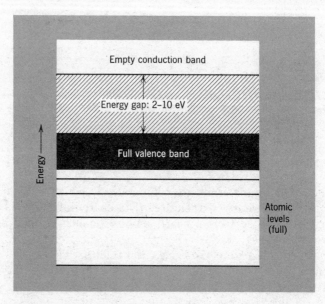

Figure 5.1   Energy-band scheme for electrons in an insulator.

it.* At ordinary voltages, very little current will flow, hence, the material behaves as an insulator. There are, however, other ways for electrons to cross the energy gap. The energy of thermal motion will allow a small fraction of the electrons in the valence band to enter the conduction band. This fraction increases with increasing temperature, as the Fermi-Dirac distribution indicates. Indeed, the electrical conductivity of an insulator increases as temperature rises. Furthermore, Figure 5.1 only applies to pure, structurally perfect insulators. Impurities and imperfections can alter the situation by creating new energy levels which increase the possibility of transitions between bands.

If the energy gap is narrow enough, a large number of electrons may be thermally excited from the valence to the conduction band. The electrical conductivity of such a material will lie between that of a conductor and an insulator. Each electron that

---

* See footnote to Section 4.2. When the field is sufficiently high, the insulator will conduct electricity, often altering its characteristic properties. This phenomenon called dielectric breakdown is discussed in chapter 12.

moves to the conduction band leaves behind a vacant state, or hole, in the valence band.   The holes behave like particles of positive charge.   Thus, each thermal excitation of an electron to the conduction band liberates not one, but two charge carriers.   When this occurs often enough, considerable conduction can occur, and the material is then called an *intrinsic semiconductor* rather than an insulator.   The mechanism of conduction is basically the same in both cases.   Customarily (and arbitrarily), materials in which the gap is less than 2 *eV* are called intrinsic semiconductors.

5.2   VALENCE-BOND MODEL

Intrinsic semiconductors, such as pure germanium and silicon, have the cubic structure of diamond, described in Volume I, Chapter 3.   The bonding in this structure is predominately covalent in character, and therefore highly directional.   The bonding electrons may be regarded as being localized between the atoms along the bonding direction.   In a structurally perfect, ideally covalent solid, the electrons are restricted to the bonds. They are unable to move through the crystal unless enough thermal energy is available to break them away from their bonding state to a free state.   In the language of the band model, one can say that an electron has to be promoted from the valence (bonding) band to the conduction band.   The situation is illustrated in Figure 5.2 for a two-dimensional cubic crystal.   The hole (vacant state in an electron pair bond) is also free to move, and acts as a carrier of positive charge, although it is, in reality, an absence of a negative charge.

The concept of a hole in the valence band is, in many ways similar, to that of a vacancy in a crystal (discussed in Section 4.1 of Volume I).   The motion of a hole to the right is equivalent to the motion of an electron to the left, since negative charge has been transferred to the left.   During the motion of the vacancy, the atomic motions are relatively simple.   The cooperative motions of the valence electrons referred to collectively as the motion of a hole, are not.   On this account it is preferable to consider the motions of a fictitious particle, the hole, rather than the more complicated collective motion of all the remaining electrons in the valence band.

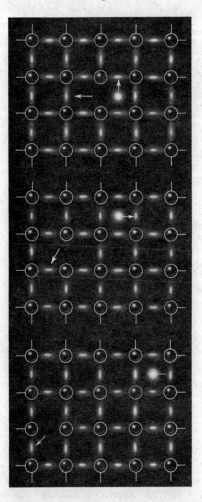

(a) An electron breaks away from the covalent bond, leaving a vacant bonding state, or a hole. The electron is now free to move in an electric field. In terms of the band model, the electron has gone from the valence band to the conduction band, leaving a hole in the valence band. The electron is shown moving upward, and the hole to the left.

(b) The conduction electron will now move to the right; the hole, down and to the left.

(c) The motions of (b) have been completed; the hole and electron continue to move outward.

Figure 5.2   Creation and motion of a conduction electron and hole in an insulator or semiconductor.

In an intrinsic semiconductor the number of holes is equal to the number of conduction electrons. When an electric field is applied, the electrons move in one direction and the holes in the opposite direction.

5.3   MODEL OF HOLE CONDUCTION

The concept of electrical conduction by means of holes, or vacant states, in the valence band follows from the wave mechanical model of conductivity described in Section 4.3. In intrinsic semiconductors, thermal energy can excite relatively few electrons into the conduction band. This process leaves a small number of vacant states at the top of the valence band. Bonding electrons near the top of the valence band can thus participate, to a limited extent, in the conduction process.

We have seen in Figure 4.4 that the curvature $d^2E/d\kappa^2$ of the E versus κ curve becomes negative at the top of the valence band. Therefore, these electrons are accelerated according to Equation 4.14:

$$a = e\mathcal{E}\,\frac{4\pi^2}{h^2}\frac{d^2E}{d\kappa^2} = \frac{e\mathcal{E}}{m^*}$$

in a direction opposite to the applied field $\mathcal{E}$. In Chapter 4, we assigned a *negative effective mass m^** to these states that we are here, for the sake of brevity, calling *holes*.

Alternatively, from Equation 4.14, we can look upon the hole as having a positive mass if we reverse the sign of the charge e and make it positive also. The convention of ascribing hole conduction to positively charged carriers grew out of early experiments, particularly the Hall effect, which first indicated the existence of holes. We should realize, however, that the motion of electrons of negative effective mass, and the motion of holes (of positive mass) are really two ways of looking at the same situation namely, the manner in which vacant states in a nearly full band allow the electrons in the band to carry electrical current. Both points of view are neat ways to bury a lot of difficult quantum mechanics in naive, yet workable fashion.

5.4   EXTRINSIC SEMICONDUCTORS

Let us now examine the effect of certain impurity atoms on the carrier concentration, using a two-dimensional valence-bond model. In the case of Ge and Si, the valence of the semiconductor is four. The square lattice of Figure 5.3 may therefore be viewed

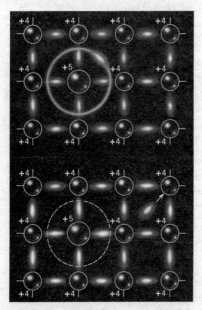

(a) A pentavalent impurity atom, with full donor state. The extra electron orbits about the donor atom.

(b) The electron has left the donor state and entered the conduction band. The empty donor state is left behind, anchored to the impurity atom.

Figure 5.3   Valence-bond representation of a donor impurity atom and its electron. Two distortions have been made in this illustration: all of the bonds are not shown, and the radius of the donor state is much too small.

as being populated by tetravalent atoms. Suppose that we have an impurity atom with a valence of five, for instance, P, As or Sb dissolved substitutionally in Si or Ge. Only four of the five electrons of the impurity atom can participate in the bonding, since there are only four bonds. Thus, four states in the valence band are available. The fifth electron does not enter a bonding state. Rather than going to the conduction band, it is attracted to the positively charged region of the impurity atom. A series of quantum states can be assigned to this additional electron similar to those of the single electron in the hydrogen atom. The binding energy however is much less—of the order of $0.01\ eV$. The binding energy is low because the region around the charged impurity atom is polarized by the electrical field of the atom. The polarization neutralizes most of the field, and the attraction for the extra electron is therefore quite weak. Since at room temperature $kT \cong 0.025\ eV$, the free electron can readily be excited by thermal energy to the conduction band. An impurity of this kind is called a *donor*, since it donates conduction electrons without producing holes in the valence band.   A simplified band model is shown in Figure 5.4.   The *donor level* represents the ground state of the fifth electron of the impurity atom.   If pentavalent impurities exist in an otherwise intrinsic semiconductor, electrons are added to the conduction band, outnumbering the holes in the valence band. The electrons then become the *majority carriers* and the holes the *minority carriers.*   The material is now called an *n*-type extrinsic semiconductor.

Trivalent impurities substituted in the tetravalent lattice have an opposite effect.   Figures 5.5 and 5.6 portray the new situation.

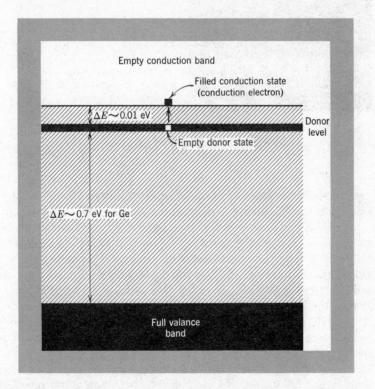

Figure 5.4   Band scheme for Figure 5.3*b*. If the conduction electron fell back into the empty donor level, Figure 5.3*a* would apply. The ΔE between the donor level and the conduction band is equal to the binding energy between the extra electron and the donor atom in Figure 5.3.

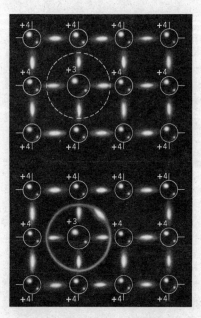

(a) A trivalent impurity atom. One bonding state must be empty due to the *missing* electron. The empty state, or *hole,* orbits about the *acceptor* atom.

(b) The hole leaves the acceptor atom, leaving a filled acceptor state behind.

Figure 5.5 Valence-bond picture of an acceptor impurity atom and the associated hole. The same distortions as those of Figure 5.3 have been made.

One of the four bonds surrounding the impurity atom is missing one electron. This leaves a hole in the valence band. If the hole moves away from the impurity atom, four saturated bonds remain as well as a net negative charge at the impurity atom. The hole is attracted to the charge and a set of quantum states similar to those of the electron in Figure 5.3 is established. The ground state of the hole is about 0.01 *eV* above the valence band. Thus the liberation of the hole from the so-called acceptor atom of Figure 5.5 is equivalent to the excitation, of an electron from the valence band to the acceptor level, as shown in Figure 5.6. A hole can be created in the valence band merely by thermal excitation. When trivalent elements such as B, Al, Ga, or In are present in otherwise pure germanium or silicon, the majority carriers are holes and the material is designated a *p*-type extrinsic semiconductor.

In many intrinsic semiconductor materials, the impurity concentration is less than one part per million. Extrinsic materials usually contain 100 to 1000 ppm. The donor impurities P, Sb, and As, and the acceptor elements B, Al, Ga, and In were mentioned first because the levels established by these elements are close enough to the conduction and valence bands, respectively, to give significant carrier concentrations at normal temperatures. Transition metals such as Fe, Ni and Co, and also Cu, when added to Si or Ge, form deep levels far removed from the band edges. Some atoms, like zinc, produce two levels rather than one. Crystal imperfections which involve broken bonds can provide localized levels. A *Frenkel imperfection* consisting of a vacancy and an interstitial, leads to some conduction. All types of levels are found near a surface. Even simple mechanical abrasion can increase the density of surface levels appreciably.

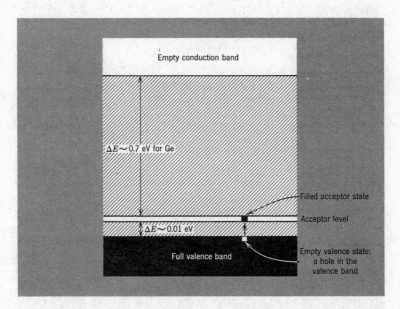

Figure 5.6 Band scheme for Figure 5.5*b*. If the electron fell back to the valence band, Figure 5.5*a* would apply. The ΔE between the acceptor level and the valence band is equal to the binding energy between the hole and the acceptor atom in Figure 5.5.

5.5   CARRIER MOBILITY

In Chapter 4 (Section 4.1), it was shown from a simple collision-acceleration model that conduction electrons do not accelerate indefinitely in an applied electric field. If they lose energy regularly through collisions, they achieve some constant average velocity, which is proportional to the applied field (see Equation 4.6). The proportionality constant between the acquired velocity (called the *drift velocity*) and the applied electric field is commonly defined as the mobility μ where

$$\mu = \frac{v_{\text{drift}}}{\mathcal{E}} \qquad (5.1)$$

It is usual to denote the mobility of conduction electrons by μ_e, and the mobility of holes by μ_h. For the metal of Section 4.1, conduction electrons flow when the field is applied. Equation 4.8 may be rewritten as

$$\sigma_e = n_e e \mu_e \qquad (5.2)$$

where n_e is the volume concentration of electrons. In the case of a semiconductor where, both electrons and holes carry current, the conductivity is given by

$$\sigma = n_e e \mu_e + n_h e \mu_h \qquad (5.3)$$

where n_h is the volume concentration of holes.

5.6   THE HALL EFFECT

Experimental verification of electrical conductivity by positive holes in semiconductors came from the Hall Effect Measurements. Consider the experiment shown in Figure 5.7. An electric field $\mathcal{E}_x$ is applied, and a current flows in the x direction. If only electrons were to flow, the current is given by Equation 5.2. The applied magnetic field exerts a force on the flowing electrons, in the z direction:

$$\mathbf{F}_Z = q\mathbf{v} \times \mathbf{B} = -eB_y v_x = -eB_y \mathcal{E}_x \mu_e \qquad (5.4)$$

The negative sign originates in the electronic charge. If no

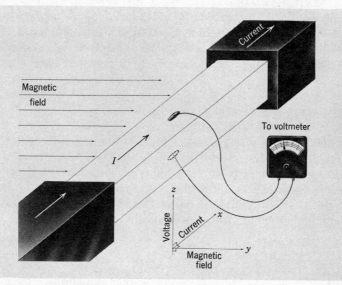

Figure 5.7 Measurement of the Hall effect. The directions of the applied magnetic field, electric current, and Hall voltage are mutually perpendicular.

current is allowed to flow out of the specimen's horizontal faces (i.e., in the z direction), the charge will simply build up until the field of the charge in the z direction cancels the force due to the magnetic field:

$$-e\mathscr{E}_z - eB_y\mathscr{E}_x\mu_e = 0 \qquad (5.5a)$$

or

$$\mathscr{E}_z = -\mathscr{E}_x B_y\mu_e = -v_x B_y \qquad (5.5b)$$

The *Hall coefficient* R_H is obtained by combining Equation 4.7 with Equation 5.5b so that

$$\mathscr{E}_H = -v_x B_y = -\frac{J_x}{n_e e}B_y = R_H J_x B_y \qquad (5.6)$$

For electrons, the Hall coefficient is negative, and the polarity of the voltmeter in Figure 5.7 can be set accordingly. This occurs in some conductors, and also *n*-type semiconductors. For *p*-type semiconductors, the Hall coefficient is positive! This result is consistent with the concept of holes as positively charged particles; the minus signs in Equations 5.5b and 5.6 then disappear. Hall conductors like Be, Zn, and Cd have positive Hall coefficients because their conduction electrons have negative mass.

5.7   EFFECT OF TEMPERATURE ON MOBILITY

The mobility of electrons or holes is influenced by scattering. The chief sources of scattering in a semiconductor are phonons and ionized impurity atoms (donors and acceptors). Other imperfections, such as dislocations, also contribute, but to a lesser degree in most useful materials. The *phonon mobility* μ_L and the ion mobility* μ_I are found theoretically to be

$$\mu_L = aT^{-3/2} \tag{5.7a}$$

$$\mu_I = bT^{3/2} \tag{5.7b}$$

where a and b are constants, for a given material. If we know the carrier electron or hole density, we can calculate the resistivity due to phonons or ionized impurities. For a single carrier, electrons for example, we can combine Equations 5.2, 5.7a, and 5.7b to obtain

$$\rho = \frac{1}{\sigma} = \frac{1}{ne\mu_L} \quad \text{or} \quad \frac{1}{ne\mu_I} \tag{5.8}$$

The total resistivity and the resultant mobility can be obtained by adding the phonon and ion resistivities, just as for conductors (Matthiessen's rule). The mobility, when both ion and phonon scatter contribute becomes

$$\frac{1}{\mu} = \frac{1}{\mu_I} + \frac{1}{\mu_L} \tag{5.9a}$$

or, with Equations 5.7a and 5.7b,

$$\mu = \frac{1}{\frac{1}{a}T^{3/2} + \frac{1}{b}T^{-3/2}} \tag{5.9b}$$

---

*Ion mobility should not be confused with ionic conduction. The term refers to carrier mobility as affected by ion scattering.

At low temperatures, the $T^{-3/2}$ term dominates, so that μ varies as $T^{3/2}$. At high temperatures, the converse holds. This signifies that ionic scattering tends to dominate at low temperatures, and phonon scattering at high temperatures, as shown in Figure 5.8. If the semiconductor is intrinsic, or nearly so, ion scattering is absent, and Equation 5.7a applies directly. This is evident in the top curve of Figure 5.8.

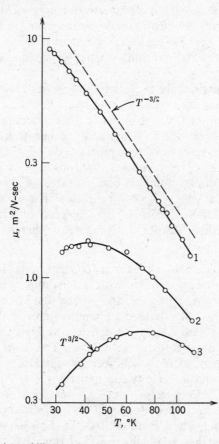

Figure 5.8   Carrier mobility as a function of temperature. Sample No. 1 is pure and therefore intrinsic, and no impurity scatter is apparent. Samples Nos. 2 and 3 are not pure, and impurity scatter leads to a $T^{3/2}$ term (Eq. 5.3b), which dominates at low temperatures, as shown by Eq. 5.9b. (After D. Long and J. Myers.)

5.8   THE EFFECT OF TEMPERATURE ON
CARRIER DENSITY AND CONDUCTIVITY

The carrier density is also sensitive to temperature since conduction electrons and holes are provided by thermal excitation. The following qualitative statements regarding the temperature variation of the carrier density in an extrinsic semiconductor then apply:

(1) At ordinary temperatures, electrons can be excited from the donor levels to the conduction band, or from the valence band to acceptor levels, because the energy required to do so is relatively small. Direct excitation from valence to conduction band is almost nil.

(2) As the temperature rises, the donor levels may become exhausted or the acceptor levels saturated, since these levels generally contain fewer states than either the conduction or valence bands (see Problem 5.2c). The carrier concentration then becomes relatively insensitive to temperature.

(3) At still higher temperatures, electrons are excited from the valence band to the conduction band in large numbers, since the sufficient thermal energy is now available. The equal quantities of electrons and holes that are liberated exceed, by far, the limited number of extrinsic carriers. Conduction, therefore, becomes intrinsic.

Figure 5.9 indicates how the carrier concentrations for an n-type extrinsic semiconductor varies with temperature. At low temperatures, statement *1* obtains, and conduction electrons are the majority carriers. At intermediate temperatures in the so called exhaustion range, statement *2* applies. At high temperatures, conduction becomes intrinsic in nature. The onset of intrinsic conduction depends on the number of states in the donor or acceptor levels and the energy gap between the valence and conduction bands. For germanium base extrinsic materials the energy gap for intrinsic conduction is $0.72\ eV$ and the upper limit for extrinsic conduction is about $100°C$. For silicon-based extrinsic materials, the gap is $1.1\ eV$, and extrinsic conduction up to $200°C$ is possible. The temperature at which intrinsic conduction becomes important sets an upper limit to the temperature at which a semiconductor device may be operated. This is described in greater detail in the next chapter.

5.9   MINORITY CARRIERS

In the extrinsic (impurity and exhaustion) ranges, the concentration of majority carriers is far greater than the minority-carrier concentration as implied in Figure 5.9.   Consider such a semiconductor when exposed to radiation by photons of energy $h\nu = \Delta E_g$. Here ΔE_g is the energy gap between the valence and conduction bands.   According to Equation 1.1 the photons are absorbed and the energy acquired raises electrons from the valence band to the conduction band.   An equal number of conduction electrons and holes is thus created.   Since the majority carrier density is large to begin with, it may not be appreciably changed, at least on a percentage basis.   The minority carriers, however, are so few in number in the unirradiated material, that the percentage increase is extremely large.   When irradiation is stopped, the excess carriers (conduction electrons and holes) simply combine and thereby disappear.   In other words, the excess conduction electrons return to

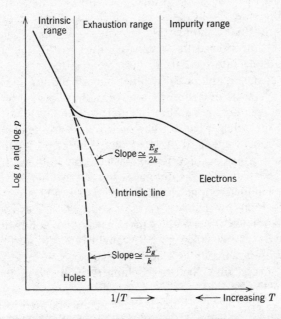

Figure 5.9   Carrier concentration as a function of temperature for an *n*-type semiconductor.

the valence band and combine with the excess holes. This process is called *recombination*. The rate of recombination is proportional to the number of excess minority carriers present:

$$\frac{dN_{ex}}{dt} = -\frac{N_{ex}}{\tau} \tag{5.10}$$

The constant of proportionality is $1/\tau$, where τ is called the minority carrier lifetime. When the irradiation stops $t = 0$ and integration of Equation 5.10 gives

$$N_{ex} = N_{ex}(0)e^{-t/\tau} \tag{5.11}$$

which describes an exponential decay of the excess minority carrier density. In the time τ, the density is reduced by a factor e^{-1}. Direct descent of conduction electrons to the valence band actually occurs relatively slowly because both momentum and energy must be conserved for each recombination. The energy of the emitted photon is fixed and its momentum is also fixed. Unless the recombining carriers have exactly the right momenta, recombination will not occur directly. However, there are usually intermediate levels available at which *phonons* can enter the process and make momentum conservation much easier. The electrons can proceed in easy steps down to the valence band. The available intermediate levels are localized about impurity atoms, vacancies, and possibly other imperfections. Thus, a conduction electron descends to a vacant, localized, intermediate level, and perhaps a number of intermediate levels, on its way down to the valence band. The final step in the sequence is recombination, or encounter with a hole. This step is facilitated by certain impurities or imperfections which *trap* both electrons and holes. They can then recombine at the intermediate levels. Impurities or imperfections are then considered to be *recombination centers*. Some local levels capture conduction electrons more easily than holes; others, holes more easily than conduction electrons. Thus, an electron may be caught in an impurity level or trap and fail to encounter a hole, or vice versa. Carriers may be trapped several times before reaching recombination centers, as shown in Figure 5.10.

Surfaces also provide imperfections with associated local levels which act as traps and recombination centers. The minority carrier lifetime is much shorter near the surface than in the in-

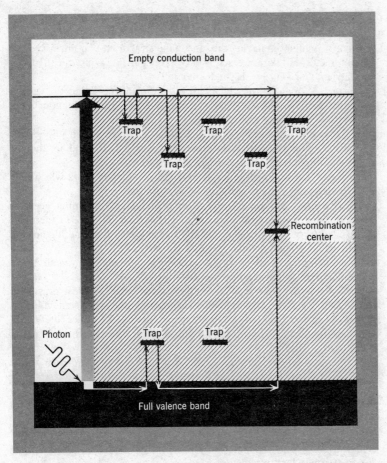

Figure 5.10   Generation, trapping, and recombination of excess minority carriers.

terior. If the surface is disturbed by abrasion, as mentioned previously, recombination becomes still greater due to the increased density of imperfections at the surface.

DEFINITIONS

Intrinsic Semiconductor. A material whose conductivity lies between those of insulators and conductors; its valence band is full, and is separated from the conduction band by an energy gap small enough to be surmounted by thermal excitation; current carriers are electrons in the conduction band and holes in the valence band in equal amounts.

Insulator.   A material having the same band structure as an intrinsic semiconductor, but with a much wider energy gap and consequently fewer carriers and very small conductivity.

Conductor.   A material in which carriers are plentiful; either the valence band is only partially filled, or the valence and conduction bands overlap.

n-Type Extrinsic Semiconductor.   A semiconductor containing donor impurities which donate electrons to the conduction band by thermal excitation; the majority carriers are electrons.

Donor Levels.   Local energy levels near the conduction band which bind the electron to the vicinity of the donor atom.

p-Type Extrinsic Semiconductor.   A semiconductor containing acceptor impurities which accommodate electrons that have been thermally excited from the full valence band in local levels; the resulting holes in the valence band are the majority carriers.

Acceptor Levels.   Local energy levels which lie near the valence band and can be occupied by excited electrons.

Majority Carriers.   The type of charge carrier which is most prevalent; holes in a p-type and conduction electrons in an n-type semiconductor.

Minority Carriers.   In an n-type semiconductor, the holes are called minority carriers; in a p-type semiconductor, the electrons are called minority carriers.

Hall Voltage.   When a magnetic field B is imposed perpendicular to the flow of current J, a voltage perpendicular to both B and J appears, called the Hall voltage.

Hall Constant or Coefficient (R_H).   The Hall voltage is proportional to B and J: $\mathcal{E}_H = R_H B J$; $\mathcal{E}_H$ is electric field, i.e., volts/meter.

Drift Velocity.   The average velocity acquired by a carrier in an applied electric field; determined by the value of the field and the amount of scattering which occurs.

Mobility (μ).   Defined by the relation $v_{(\text{drift})} = \mu \mathcal{E}$; that is, μ is the proportionality constant between the drift velocity and the applied field.

Recombination.   The process by which excess carriers produced by radiation or other means in a semiconductor crystal return to their equilibrium state when the radiation is removed; usually occurs by sequential quantum jumps called recombination transitions.

Recombination Center.   An energy level that can capture both electrons and holes.

Surface States.   Discrete energy levels produced by surface imperfections.

Trap.   An energy level that can capture either electrons or holes easily, but not both.

BIBLIOGRAPHY

Adler, R. B., Smith A. C., and Longini, R. L., *Introduction to Semiconductor Physics*, John Wiley & Sons, New York 1964, Chapters 1 and 2.

Azaroff, L. V. and Brophy, J. J., *Electronic Processes in Materials*, McGraw-Hill, New York, 1963, pp. 194–223.

Hutchison, T. S., and Baird, D. C., *The Physics of Engineering Solids*, John Wiley & Sons, New York, 1963, pp. 262–276.

Cusack, N., *The Electrical and Magnetic Properties of Solids*, John Wiley & Sons, New York 1958, pp. 192–240.

Kittel, Charles, *Elementary Solid State Physics*, John Wiley & Sons, New York, 1962, pp. 199–250.

PROBLEMS

5.1   (a) The electron configuration for sodium is $1s^2 2s^2 2p^6 3s$.   Draw a schematic energy band diagram for one mole of solid sodium.   Which levels do you expect to broaden?   Indicate (e.g., by shading) the occupation of the bands by electrons.

(b) On the diagram, indicate the number of states in each band and the number of occupied states.   Indicate the conduction and valence bands, by labeling.

(c) The electron configuration for magnesium is $1s^2 2s^2 2p^6 3s^2$, so that the 3s shell is full.   Give reasons, if you can, to explain the fact that Mg is a conductor.

(d) The configuration for carbon is $1s^2 2s^2 2p^2$.   The bonding of the carbon atoms in diamond is covalent, converting the $2s^2 2p^2$ into sp^3 *hybrid* states (see Chapter 1, Volume I) with only four hybrid levels in the shell, per atom.   Explain why diamond is an insulator.   (The coordination number for diamond is four.)

5.2   (a) Which elements would you add to pure silicon to make it a *p*-type extrinsic semiconductor?

(b) Draw a band diagram for silicon, as follows: the full valence band (four sp^3 hybrid states, filled) with the bottom of the empty conduction band 1.2 eV above the top of the full valence band.

(c) Draw a band for acceptor states, with an 0.01 eV binding energy. Suppose 1 part per million of Al is added to one mole of Si. How many states are there in the acceptor level?

5.3   Using the results of the previous problem:

(a) Calculate the carrier density in the silicon due to the aluminum impurity, at 300°K, using the Maxwell-Boltzmann distribution, that is, the probability of excitation is proportional to $e^{-\Delta E/kT}$.

(b) Calculate the intrinsic carrier density at 300°K.

(c) At what temperature is the intrinsic carrier density equal to the carrier density due to the aluminum impurity? This is the onset of the intrinsic range of Figure 5.9.

5.4   The energy gap for germanium is 0.78 eV and the binding energy for an Al acceptor level about 0.01 eV. In the same manner as for the previous problem, determine the temperature at which intrinsic processes take over.

5.5   Show that for energies far enough above the Fermi level so that $e^{(E-E_F)/kT} \gg 1$, the Fermi-Dirac and Maxwell-Boltzmann distributions are approximately the same.

5.6   How would plastic deformation affect (a) Carrier density, (b) Carrier mobility, (c) Electrical conductivity, and (d) Minority carrier lifetime of an intrinsic semiconductor?

5.7   Could the energy gap be measured by the absorption of radiation? What type of radiation? Describe a possible method, with as much experimental detail as possible.

5.8   Consider a semiconducting crystal 0.02 m wide, 0.001 m thick, with a current of 0.002 amp flowing along the length. Parallel to the long side and perpendicular to the current, a magnetic field of 0.5 webers/meter^2 is applied. The Hall constant is 3.84×10^{-4} meter^3/coulomb. Calculate the Hall voltage across the short side.

5.9   Suppose the Hall coefficient for a particular material were found to be zero. Could you explain this result?

5.10   It is possible to perform experiments of a mechanical nature to determine the mass of conduction electrons, and such experiments always yield the classical mass, even for materials where the valence band is full or nearly full and the Hall coefficient is positive. Explain.

5.11   Consider germanium, with an energy gap of 0.78 eV, with 1 part per million of Al, which has acceptor levels with binding energies of about 0.01 eV. Calculate the carrier density at 300°K, using the methods of Problem 5.3. The mobility, which is dominated by impurity scatter at 300°K, is about 0.4 meter^2/volt-second. Calculate the electrical conductivity. Suppose the carrier mobility was inversely proportional to the density of ionized impurity atoms. Could you increase the conductivity by increasing the amount of Al added to the germanium?

5.12   The electron mobility in a very pure semiconductor (Ge, for example) is 50 meter^2/volt-second at 4.2°K. What is the mobility at 300°K?

5.13   A burst of incident radiation creates 10^{11} excess minority carriers/m^3 in a semiconductor. If the material had a minority carrier lifetime of 1 microsecond, how much time was required for the excess concentration to drop 10^9/m^3?

5.14   Based on the band model of electron conductivity, what would you say about the effect of temperature on the conductivity of normal insulators, such as diamond or ionic compounds?   Check your ideas by finding some actual data on the conductivity of insulators at elevated temperatures.

5.15   Estimate the mobility for a typical metal having a resistivity of 10^{-7} ohm-m.   Estimate the mean free time τ if the effective mass of the electrons equals the classical value.

CHAPTER SIX

Semiconductor Devices

When a voltage is applied in the *forward* direction to a barrier or
p-n junction rectifier it decreases the energy barrier which
opposes the flow of majority carriers from both sides of the
junction and conduction occurs. If the voltage is applied in the
reverse direction, the barrier height is increased and very little
majority carrier flow occurs. The energy barrier is due to the
contact potential arising from the equalization of the Fermi
energies of the materials in contact. There is no barrier at a *p-n
junction* to the flow of minority carriers from either side of the
junction; however, usually minority carriers are sufficiently rare
that only majority-carrier flow is useful. However, minority
carriers may be injected into the material near the junction, and
large increases in the flow of current through the junction will then
occur, even if the junction is *reverse-biased.* If the material on one
side of the junction contains another junction nearby, the second
junction may be *forward-biased* and used for the controlled injection of minority carriers. The *transistor* is a double-junction
device. Photocells use radiation-induced minority carrier injection. Photoconductors and thermistors operate on the increase
in carrier density (and, therefore, conductivity) that results from
radiation and increasing temperature, respectively. The tunnel
diode functions as a result of quantum mechanical tunneling
through a thin junction. Because the *p-n* junction—the really
essential part of a semiconductor device—is small in size, extreme
miniaturization of semiconductor circuits is possible.

6.1 INTRODUCTION

This chapter is primarily concerned with the basic operation of
the *p-n rectifying junction, the transistor,* and the *tunnel diode.* The

116

present section deals with necessary definitions, conventions, and symbols commonly used in the description of such devices. The current versus voltage plot shown in Figure 6.1 for an ideal rectifier provides a convenient starting point. An ideal rectifier conducts the electrical current in one direction only. Its resistance in the opposite direction may therefore be considered infinite. It is conventional to view the electrical current as a flow of positive charge. The *rectifier* shown in Figure 6.1 will conduct the current when the potential at the left-hand connection is higher than at the right. This situation is called *forward bias*. If the voltage at the right-hand connection is higher, the rectifier is said to be on *reverse bias;* very little current then flows. The flow of current from left to right is indicated by an arrow on the left and a gate on the right.

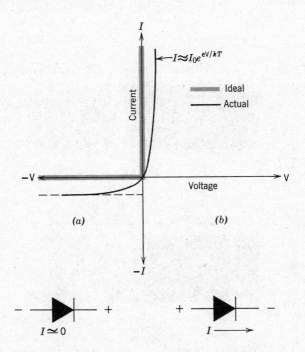

Figure 6.1 Electrical characteristics of an ideal rectifier. Resistance for positive voltage is low; for negative voltage, high. (*a*) Negative (or reverse) bias voltage. (*b*) Positive (or forward) bias voltage.

The boundary between *p*- and *n*-type extrinsic semiconductor material is called a *p-n junction*. Such a junction can be used as a rectifier. A conventional current (positive charge) can flow from the *p*-type side of the junction to the *n*-type side, but not in the opposite direction. The term junction is deceptive. It is difficult to construct a *p-n* junction by simply joining pieces of *p*- and *n*-type material together without introducing surface imperfections and impurities which obscure rectifying action. Practical rectifiers containing adjacent *p*- and *n*-type regions are therefore carefully made by controlling the impurity content of single crystals (Ge and Si). Indeed, any useful junction whether in a rectifier, transistor or other devices should occur in a single crystal. Figure 6.2 shows such a single piece of rectifier material with the left-side *p*-type extrinsic and the right side *n*-type. If this rectifier is reverse biased, the majority carriers are quickly exhausted. The rectifier then becomes polarized. If the rectifier is forward biased,

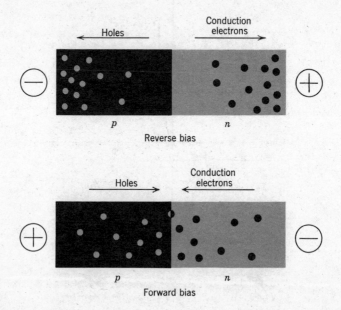

Figure 6.2   Valence-bond picture of *p-n* junction rectification. Biased in reverse, the rectifier polarizes, and little current passes. Forward biasing, with continual annihilation of conduction electrons near the junction by recombination with holes, allows large currents to pass.

excess majority carriers enter the *p* and *n* regions. They can recombine continuously at or near the junction. Recombination need not occur right at the junction for conduction electrons coming from the *n*-type side may penetrate deeply in the *p*-type material before recombining. The converse applies to holes originating in the *p*-type material. In order to understand how a *p-n* junction functions as a rectifier, it is useful to begin with a discussion of barrier rectification and contact potential in general.

6.2   BARRIER RECTIFICATION AND CONTACT POTENTIAL

Because electrons flow with ease only from the filament to the plate in a vacuum diode it has long been used as a rectifier. A number of semiconducting rectifiers were also used for many years prior to the introduction of the *p-n* junction rectifier. In the *dry disc rectifier,* a copper plate is separated from a lead plate by a layer of nonconducting cuprous oxide. The selenium rectifier, consists of a layer of semiconducting selenium separated from a metal layer by an insulating layer (usually *amorphous* selenium). In both cases, two conducting or semiconducting materials are separated by an insulating or barrier layer. To understand why rectification takes place, it is well to visualize what happens when the two materials are brought into contact with an insulator. Figure 6.3 depicts the situation for two metals. The behavior near the surface as described in Chapter 2 is not shown. Before contact, the two metals are at the same potential. Their *vacuum state* has then the same energy. Since their Fermi levels are at different energies, an electron can lower its energy by dropping as indicated in Figure 6.3, from the Fermi level of Metal 2 to that of Metal 1. On coming to equilibrium after contact is made the electrons therefore gradually flow through the insulating barrier from Metal 2 to Metal 1.

Two effects tend to match up the Fermi levels. The first includes the emptying of states in Metal 2 and the filling of states in Metal 1. The second concerns the potential difference arising from the negative charge taken on by Metal 1 and the positive charge taken on by Metal 2, due to the transfer of electrons. Of the two, the charge transfer has much the greater effect. When about one electron in 10^{15} in the conduction band has been transferred, a potential difference of the order of one volt is established. Let us therefore ignore the emptying and filling of states. The lower part (after contact) of Figure 6.3 shows Metal 1 negatively charged, and the energy band diagram for Metal 1 moved up a distance $e(\phi_1 - \phi_2)$ relative to the diagram for Metal 2. The difference in potential $(\phi_1 - \phi_2)$ is called the *contact potential*. It is about one volt in magnitude and may be obtained from the

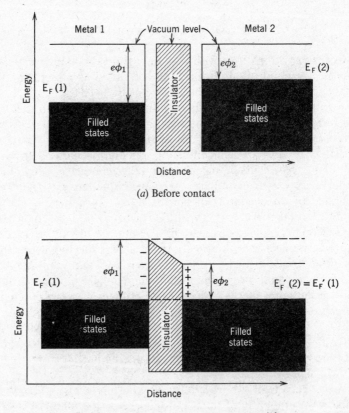

(a) Before contact

(b) After contact; $\phi_1 - \phi_2 \equiv$ contact potential

Figure 6.3 The effect of contact on two conductors, with an insulating layer between. Electrons flow from Metal 2, which has the higher Fermi level, to Metal 1, until the charge difference raises the potential of Metal 1 enough to match the Fermi levels, bringing the metals to equilibrium.

work functions of the two metals. The contact potential is not generally measured with a voltmeter for the voltage difference between the two metals in equilibrium in Figure 6.3b is zero. We can, however, measure the charge transfer which occurs when the metals are brought into contact.

On an intuitive basis, we should expect electrons to proceed from the right to the left in Figure 6.3 more easily than in the op-

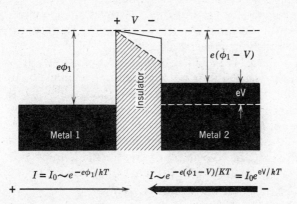

$$I = I_0 \sim e^{-e\phi_1/kT} \qquad I \sim e^{-e(\phi_1 - V)/KT} = I_0 e^{eV/kT}$$

(a) Forward bias. Metal 2 is made negative, and a large flow of electrons to the left occurs

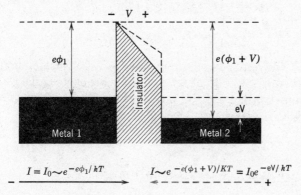

$$I = I_0 \sim e^{-e\phi_1/kT} \qquad I \sim e^{-e(\phi_1 + V)/KT} = I_0 e^{-eV/kT}$$

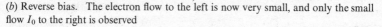

(b) Reverse bias. The electron flow to the left is now very small, and only the small flow I_0 to the right is observed

Figure 6.4   Rectification by the barrier and junction of Figure 6.3.

posite direction. This is true, for the low-resistance direction of
electron flow is always from the metal of low work function to the
metal of high work function. Figure 6.4 shows the *barrier rectifier*
of Figure 6.3 with forward- and reverse-bias voltages. The electrons on the left face the barrier $e\phi_1$, which is not changed by the
applied voltage. If the barrier is too thick for tunneling or
thermal excitation (or some other source of energy) it is necessary
that the flow of electrons pass the barrier from left to right. The
probability for doing so is proportional to $e^{-e\phi/kT}$. The situation
may be compared to the thermal electron emission phenomenon
discussed in Chapter 2. It is however necessary to employ
the Maxwell-Boltzmann exponential $e^{-e\phi/kT}$ in this case, rather
than the Fermi-Dirac distribution. As suggested in Problem
5.5, the Fermi-Dirac and Maxwell-Boltzmann functions are
essentially identical for energies more than a few kT (at room
temperature, $kT \sim 1/40\ eV$) above the Fermi energy. The
applied voltage indicated in Figure 6.3 does not affect the thermally excited electron flow from left to right. It does, however,
shift the barrier faced by electrons on the right, in Metal 2. When
the barrier is forward-biased (Figure 6.4a), a large net current to
the left results. Its magnitude is given by

$$I(\text{net}) = \overleftarrow{I} - \overrightarrow{I} = I_0(e^{eV/kT} - 1) \approx I_0 e^{eV/kT} \qquad (6.1a)$$

When the barrier is reverse-biased (Figure 6.4b), the shift in the
barrier gives a net current

$$I(\text{net}) = I_0(e^{-eV/kT} - 1) \approx -I_0 \qquad (6.1b)$$

(One volt corresponds to about $40kT/e$ at room temperature.)
Equations 6.1a and 6.1b also account for the exponential forward
current and constant reverse current of the ideal rectifier shown in
Figure 6.1.

6.3   FERMI LEVEL IN SEMICONDUCTORS

Let us now examine the *p-n* junction in the same manner as the
barrier rectifier. To locate the Fermi level in a semiconductor, the
Fermi function

$$f(E) = \frac{1}{1 + e^{(E-E_F)/kT}} \qquad (6.2)$$

Since its required derivation does not depend on the type of band structure involved, the occupation of the states in a semiconductor should follow Equation 6.2. It is, however, necessary to remember that no states exist in the forbidden bands. We may therefore superimpose the Fermi-Dirac distribution on the band structure to get the actual energy distribution of the electrons. Figure 6.5 shows the Fermi-Dirac distribution and the filling of states in an intrinsic semiconductor. The Fermi level is located in the middle of the energy gap, so that the number of electrons in the conduction band is equal to the number of holes in the valence band.

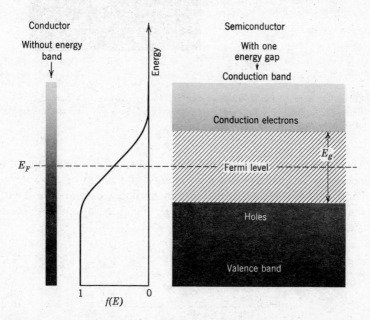

Figure 6.5   The occupation of states in a perfect conductor (*left*, narrow bar) and an intrinsic semiconductor (*right*, broad band diagram), having the same Fermi level, at temperatures well above zero °K. In the center is the Fermi-Dirac distribution function, which gives the occupation probability, or fraction of available states occupied. The depth of the shading is proportional to the fraction of states occupied, i.e., the value of the Fermi-Dirac function. The effect of the forbidden band of the semiconductor is to wipe out one section of the distribution.

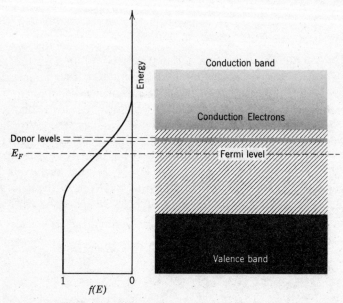

Figure 6.6   The occupation of states in an n-type extrinsic semiconductor. Intensity of shading is proportional to the fraction of states occupied, which is given by the Fermi-Dirac function $f(E)$. The Fermi level is now shifted upward from the middle of the gap, due to the extra electrons present.

The Fermi level can also be found algebraically. To do this it is necessary to place the zero of energy at the top of the valence band, and equate the probability of finding a hole at $E = 0$ with that of finding an electron at $E = Eg$ (i.e., the bottom of the conduction band) where

$$1 - f(0) = f(E_g),\tag{6.3}$$

by employing Equation 6.2,

$$1 - \frac{1}{1 + e^{-E_F/kT}} = \frac{1}{1 + e^{(E_g - E_F)/kT}}\tag{6.4}$$

and, solving for E_F,

$$E_F = \frac{E_g}{2}\tag{6.5}$$

The above calculations involve two assumptions: (1) all the holes

are at the very top of the valence band and all the conduction electrons are at the very bottom of the conduction band; and (2) the
density of states does not influence the answer.   The Fermi
function may thus be applied directly to a semiconductor.   The
presence of the forbidden energy band merely eliminates the
forbidden energies from the actual distribution of electrons, as
shown in Figure 6.5.

In an *n*-type extrinsic semiconductor, the electrons from ionized
donors can fill the holes in the valence band as well as enter the
conduction band.   This shifts the whole distribution upward, as
shown in Figure 6.6.   If the donor atom concentration is increased
the Fermi level is raised.   In a *p*-type material however the
acceptor band can accommodate electrons from the conduction
band and valence band.   This shifts the Fermi level downward, as
shown in Figure 6.7.

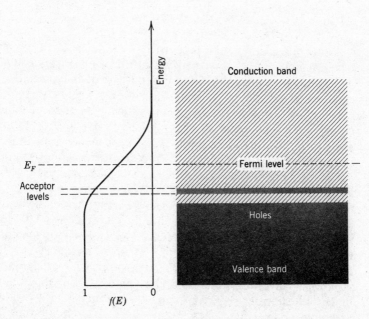

Figure 6.7   The occupation of states in a *p*-type extrinsic semiconductor.   Intensity
of shading is proportional to the fraction of states occupied, which is given by the
Fermi-Dirac function $f(E)$.   The Fermi level is now shifted downward, due to the
availability of acceptor levels for the electrons.

Since an extrinsic semiconductor becomes intrinsic as the temperature is raised, the Fermi level moves toward the middle of the energy gap as in Figures 6.6 and 6.7. Furthermore the filling of donor and acceptor levels must also obey the Fermi-Dirac distribution. This means that the *exhaustion range* discussed in Chapter 5 does not correspond to the actual emptying of all donor levels, or complete filling of all acceptor levels. What actually happens as the temperature rises is that the Fermi level moves away from the donor or acceptor levels toward the middle of the energy gap. Even before intrinsic behavior sets in, the separation of the Fermi level from the acceptor or donor band gets so large that the ionization probability is no longer very sensitive to temperature because fewer electrons occupy these states. The carrier density therefore only becomes temperature-sensitive when the temperature for intrinsic behavior is reached.

6.4   *p-n* JUNCTION RECTIFICATION

Let us now examine the properties of the boundary between a *p*-type region and an *n*-type region in the same single crystal of semiconducting material. For this purpose it is convenient to choose a sharp boundary, contained within a single crystal. If a *p-n junction* is fabricated by some particular method of joining *p*-type and *n*-type material, or if a grain of *p*-type merely abuts a grain of *n*-type, the extra levels found near the surfaces or grain boundaries can so complicate the problem that rectification does not occur. Figure 6.8 depicts the energy bands of a single crystal *p-n* junction. Because the Fermi level is lower on the *p*-side relative to the conduction band, electrons move across the boundary to the *p*-side and thereby equalize the Fermi levels. This gives rise to the contact potential difference V_0. Holes are the majority carriers on the left and conduction electrons on the right. It is well to keep in mind that Figure 6.8 is an energy level diagram for electrons. Inverting Figure 6.8 gives an analogous diagram for holes. Electrons in Figure 6.8 can lower their energies by sinking and holes lower their energies by rising. At equilibrium, the minority carriers lower their energies by crossing the junction; electrons pass to the right and holes to the left. Some time after crossing, the electrons and holes recombine, as discussed in

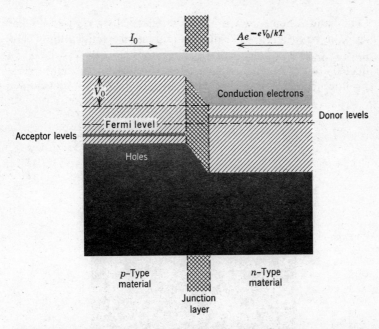

Figure 6.8   Energy band diagram for a *p-n* junction rectifier. The contact potential V_0 is shown. The filling of the states is designated by the shading. The flow of minority electrons from the *p* region is equal to the number of majority electrons from the *n* region surmounting the barrier; thus, the net current is zero.

Chapter 5.   Because the minority carrier concentrations are usually small, such currents are expected to be small.

The flow of electrons (minority carriers from the *p* side) to the right in Figure 6.8 must be balanced at equilibrium by an equal flow to the left. This means that the flow of minority carriers from the *p*-type material is balanced by a flow of majority carriers (electrons again) from the *n*-type material. In doing so, the conduction electrons in the *n*-type material face the contact potential barrier V_0. The crossing probability is then proportional to $e^{-eV_0/kT}$. If the flow to the left is $Ae^{-eV_0/kT}$, the current at equilibrium is

$$I_0 = Ae^{-eV_0/kT} \tag{6.6}$$

A similar expression may be derived for the flow of holes from both sides.

The band scheme shown in Figure 6.9 for forward and reverse bias can be used to visualize *p-n* junction rectification. The minority current of electrons, from the *p* side to the *n* side, is not affected for no barrier to flow from the left to the right exists. The flow from the *n* side is very sensitive to the applied voltage

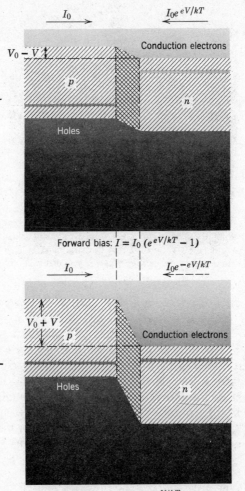

Forward bias: $I = I_0 (e^{eV/kT} - 1)$

Reverse bias: $I = I_0 (e^{-eV/kT} - 1)$

Figure 6.9   The junction rectifier of Figure 6.8, with forward and reverse bias.

for the electrons in the n-type material face a barrier which may be lowered or raised by the applied voltage. Thus, for reverse bias, the majority flow is $Ae^{-e(V_0+V)/kT}$, or $I_0e^{-eV/kT}$, where V is the applied voltage. The net current is the sum of two currents: 1) the minority flow from the left, and 2) the majority flow from the right. It is small and equal to $I_0(e^{-eV/kT} - 1)$. For forward bias, the majority flow is $Ae^{-(V_0-V)/kT}$, or $I_0e^{eV/kT}$. The net flow in the opposite direction is $I_0(e^{-eV/kT} - 1)$ which increases exponentially with voltage. This gives the same current versus voltage relationship illustrated in Figure 6.1.

The rectifying action of the junction, for electrons described above can also be demonstrated for holes. It is necessary however to point out two important limitations to rectifying action in this case. The diode consists of a crystal, usually silicon or germanium, with acceptor impurities on one side of the junction and donor impurities on the other side. The barrier and rectification effect arise from the difference in the Fermi levels. If the temperature is high enough, both sides will be intrinsic. Since both Fermi levels will then be at the center of the gap, the contact potential vanishes, and rectification ceases. The other limitation involves the injection of excess minority carriers into the junction region, by radiation, for example. The excess carriers will then increase the net current, especially if the junction is on reverse bias.

6.5   TRANSISTORS

In 1947, Bardeen and Brattain found that a device consisting of two closely spaced points on a crystal of semiconductor material could amplify a signal. The discovery of their point-contact transistor was soon followed by the junction transistor of Shockley. It is essentially a single crystal with two p-n junctions arranged back to back. In the p-n-p type shown in Figure 6.9 the two p regions are separated by an n region. The n-p-n transistor has two n regions separated by a p region. In Figure 6.10, junction 1 is forward biased. The p region of junction 1 is called the *emitter,* and the n region the *base.* Large numbers of electrons and holes flow across junction 1; electrons from right to left, and holes in the opposite direction. In the p-n-p transistor, the holes which are injected into the base from the emitter are important. In the case

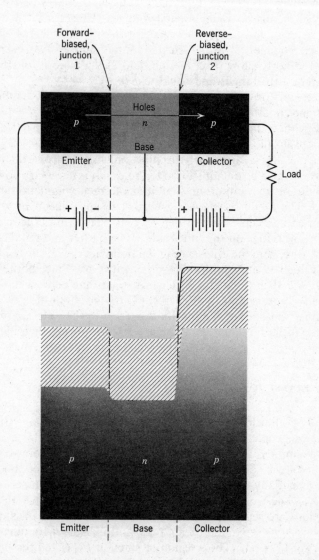

Figure 6.10   A *p-n-p* transistor.   Holes pass through junction 1 to appear as minority carriers at junction 2.

of a rectifier, we should simply expect the injected holes which are excess minority carriers in the *n*-type base to recombine. In the present case, let us assume that the base is thin enough and the minority carrier lifetime in the base long enough so that the holes penetrate the base without recombination and arrive at junction *2*. Junction *2* is reverse-biased, and holes are the minority carriers in the base. Since there is no barrier at the junction for the holes, and the *equilibrium* concentration of holes in *n*-type material is quite small, the hole current on reverse bias must be small. The *excess* holes injected by junction *1* can increase the reverse current of junction *2* sharply. In a good transistor, almost all of the holes generated at junction *1* are injected as excess minority carriers into junction *2*. The reverse current through junction *2* is then almost equal to the forward current through junction *1*. The current through junction *2* can be controlled simply by controlling the current through junction *1*, that is, by controlling the hole current through the transistor.

To operate the transistor efficiently, requires hole generation at junction *1*, minimal losses by recombination in the base, and efficient hole extraction at junction *2*. Although junction *1* is forward-biased and may be carrying a large current, the current is largely due to electrons flowing from the base to the emitter, and therefore largely useless. What is needed is a hole current flowing to the base. The ratio of emitter hole current to total emitter current is called the *injection efficiency* (means for obtaining greater injection efficiency are discussed in Chapter 7). To minimize losses of injected holes in the base, the base must be made as thin as possible and the minority carrier lifetime increased. This is also described in Chapter 7.

Suppose the voltage applied to junction *2* is quite large compared to the voltage across junction *1*. As long as junction *2* is reverse-biased, it will collect holes from the base, independent of the voltage across it. A small increase in voltage across junction 1 will give a large increase in the hole current injected into the base which, in turn, means a large increase in the current across junction *2*. The large increase in collector current is reflected by a large voltage increase across the load resistor in Figure 6.10. Thus a small voltage change in the emitter circuit causes a large one in the collector circuit. The load resistor and the applied voltage if made large, increase the amplification because the reverse current in the base-collector junction is not sensitive to the voltage across it. It is also possible to amplify a current, for small changes in the current to the base result in large changes in the collector current.

Similar reasoning may be applied to describe an *n-p-n* transistor. Here conduction electrons are injected by the *n*-type emitter into the *p*-type base, and are collected at the base-collector junction. In both kinds of transistors high temperature and irradiation can disrupt operation in the same manner as they disrupt rectification. As mentioned previously, this is caused by the onset of intrinsic behaviour and excess carrier injection.

6.6   TUNNEL DIODE

As stated in Chapter 2, electrons may pass through thin enough barriers without change in energy by tunneling. For tunneling to occur the Pauli Exclusion Principle requires the presence of vacant states of the same energy on the other side of the barrier. Figure 6.11 illustrates the energy band diagram of a semiconducting device based on tunneling. Both the *p* region and the *n* region are heavily loaded with acceptor and donor impurities, respectively. As a result, the Fermi levels lie in the valence band and conduction band, respectively. A very large contact potential exists because of the wide separation of Fermi levels. This makes normal barrier crossing impossible until a considerable voltage is applied. The junction width is made as small as possible to promote tunneling. The top of Figure 6.11 shows the band scheme of the device at zero bias. The middle of the figure indicates the effect of a moderate forward bias. If the electrons in the *n* region tunnel to the empty states *of the same energy* in the *p* region, current will flow. Further increase in bias will markedly lower this current as indicated in the bottom diagram of Figure 6.11. Electrons in the *n* region will find only the forbidden band and full states facing them on the other side of the junction; very little current then flows. The current actually decreases as the voltage is increased. If still higher voltages are applied to the diode, the barrier is lowered enough for ordinary rectification processes to occur.

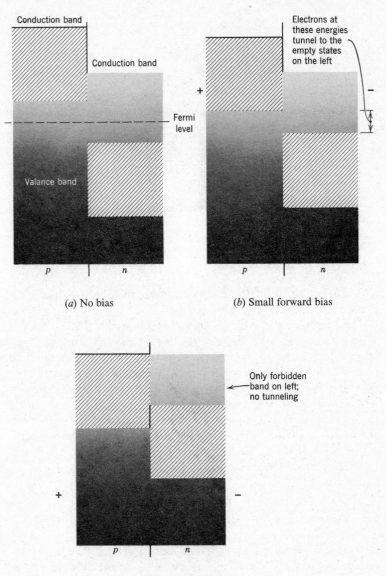

(a) No bias

(b) Small forward bias

(c) Larger forward bias

Figure 6.11   Band scheme for the Esaki diode.   Tunneling current is large only when empty, and full states are at equal energies across the barrier, as in (b).

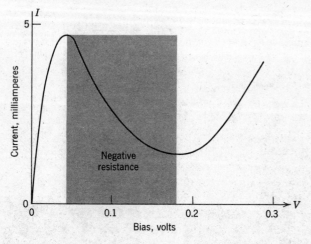

Figure 6.12   Current versus voltage curve for a typical Esaki diode.

The current versus voltage curve is then substantially that of a junction rectifier.

The current versus voltage curve for a tunnel or Esaki diode is shown in Figure 6.12. The region of decreasing current corresponds to a negative resistance. Normal circuit components in contrast have a positive resistance. This means that we may consider the diode when operated in the negative resistance region as a power amplifier or source. When biased in this region and used in conjunction with a capacitor and an inductance, the circuit can act as an oscillator. Such circuits may be used to generate frequencies as high as 10^{11} cps because the tunneling process is essentially instantaneous.

6.7 PHOTOCELLS AND PHOTOCONDUCTORS

Transitions induced by radiation can also be utilized in semiconducting devices. Radiation may be absorbed by electrons in the valence band if the photons are energetic enough to excite the electrons into empty states. In many cases this means excitation across the energy gap into the conduction band. The condition for excitation then is $h\nu \geqslant \Delta E_g$. The creation of additional carriers increases the conductivity. If the yield of carriers per incident photon is high enough, and the excess carrier lifetime large enough, the *photoconductive* material may be used to measure the intensity of radiation.

If radiation creates electron-hole pairs near a reverse-biased *p-n* junction, the current increases sharply, due to the increase in minority carrier density. The situation is similar to the minority carrier injection effect in transistors. There are no energy barriers at the junction to the flow of minority carriers, and the reverse current therefore rises sharply. The reverse current will depend on the rate of injection of minority carriers, which depends on the intensity of the incident radiation. Thus, the current in the reverse-biased junction may be used to measure the intensity of light. If the bias on the junction is zero, the minority carriers will still cross the junction to lower their energies. Continued radiation will cause more electrons and holes to pile up at opposite sides of the junction, and a voltage will be generated. This will cause a current to flow in a connected circuit. The energy of the incident photons may thus be converted to electrical energy by excitation of electron-hole pairs. If the junction responds to visible or near-visible light, the device is called a *solar battery*.

6.8   THERMISTORS

Since carriers are created by thermal excitation, the resistivity of a semiconductor changes with temperature. Semiconductors used in temperature measuring devices are called thermistors. For this purpose, polycrystalline oxide ceramics such as MnO containing dissolved lithium ions, magnetite (Fe_3O_4), or extrinsic single crystal semiconductor (Ge) are useful. Thermistors capable of measuring temperatures up to about 450°C and others suitable down to 1°K are commercially available. Such materials have sensitivities of millidegrees; hence precision auxiliary equipment is not necessary if measurements only require 0.1°K accuracy. Thermistors are also employed to advantage as temperature compensators in electronic circuitry.

6.9 MICROELECTRONICS

The necessities of the space age put a premium on devices of small size and small power requirements. In this connection, semiconductors are ideal; the only essential parts of a transistor are really the two junctions, and the efficiency can go up as the size goes down. Many techniques are used to reduce the size of a circuit: tiny components, *printed* circuits, *thin-film circuits,* and *three-dimensional* circuitry. The ultimate appears to be the "integrated circuit" approach, where the entire circuit is fabricated from a single crystal of a semiconductor material, and different regions of the crystal act as diodes, transistors, resistors, and capacitors. The connections between the different regions are high-conductivity paths in the crystal. Figure 6.13 shows such microcircuits. There are 40 complete *circuits* on the wafer, each containing the equivalent of many transistors, diodes, resistors,

Figure 6.13   Microcircuits on silicon wafer (Courtesy Western Electric Co.).

and capacitors. Besides difficult fabrication the lower limit to component size is imposed by heat evolution and cosmic radiation, both of which grow more serious as size decreases.

DEFINITIONS

Rectifier (Diode).   An electronic device which, conducts electric current in one direction only; in the other direction, the device is, an insulator.

Barrier Rectification.   Rectification due to the presence of an insulating layer between two dissimilar conducting materials.

Bias.   The voltage applied to a rectifying junction or any other two electrodes of an electronic device.

Forward Bias.   Bias applied to a rectifying junction in the conducting direction.

Reverse Bias.   Bias applied in the insulating direction to a rectifying junction.

Contact Potential.   The potential difference of two solids due to the difference in the Fermi levels. Equal to the difference in work function of the two solids.

Junction Rectification.   Rectifying action at a *p-n* junction.

p-n Junction.   An abrupt boundary, in a single crystal of semiconducting material between *n*-type and *p*-type regions.

Transistor.   A two-junction device. One junction is forward-biased, injecting minority carriers into the base. The second junction is reverse-biased to a relatively large voltage. The current through the second junction is controlled by the current through the first.

Injection Efficiency.   The ratio of useful carriers to total carriers passing the emitter-base junction in a transistor. By useful, we mean carriers which are in the minority in the base region.

Base.   The middle section of an *n-p-n* or *p-n-p* transistor biased for amplification.

Emitter.   In a transistor, the emitter-base junction is forward-biased.

Collector.   In a transistor, the collector-base junction is reverse-biased, and minority carriers are injected into the collector-base junction by the nearby emitter-base junction.

Tunnel (Esaki) Diode.   A very narrow *p-n* junction of materials in which the carrier density is extremely high, in which carriers cross the junction by tunneling. A negative resistance range in the I versus V curve results.

Photoconductor.   A semiconductor whose conductivity is sensitive to light due to the creation of conduction electron-hole pairs by absorption of photons.

Photocell. A junction whose sensitivity to light is due to the extra minority
carriers excited by the radiation.

Thermistor. A semiconductor thermometer, which uses the temperature
dependence of the carrier densities.

Microelectronics. The extreme miniaturization of circuits. In its most
extreme form, the integration of circuits, junctions, etc., into a single,
small chip of solid material.

BIBLIOGRAPHY

ELEMENTARY READING:

Holden, Alan, *Conductors and Semiconductors,* Bell Telephone Labs, 1964.

SUPPLEMENTARY READING:

Azaroff, L. V., and Brophy, J. J., *Electronic Processes in Materials,* McGraw-Hill, New
York, 1963, Chapters 8 and 10.
Sproull, R. L., *Modern Physics,* Wiley, New York, 1963, Second Edition, Chapter 11.
Van der Ziel, A., *Solid State Physical Electronics,* Prentice Hall, New York, 1957.

ADVANCED READING:

Cusack, N., *The Electrical and Magnetic Properties of Solids,* Wiley, New York, 1958
(Second Printing, 1960), pp. 39–54, and Chapter 10.
Kittel, C., *Introduction to Solid State Physics,* Wiley, New York, 1960, Chapter 14.

PROBLEMS

6.1 What is the effect of temperature on a barrier or junction rectifier?
Refer to Equations 6.1 and 6.4. Use *I-V* plots, and show the changes due
to temperature.

6.2 Why are extrinsic materials needed for junction rectifiers? Could
intrinsic semiconductors do the job?

6.3 Consider a metal-semiconductor junction. Is rectification possible?
Draw the energy band diagram for a metal *p*-type semiconductor junction
and explain.

6.4 Is rectification possible in a metal *n*-type semiconductor junction?
Draw the energy band diagram, and explain. How about a metal intrinsic
semiconductor junction?

6.5 Is junction rectification possible at the junction between two conductors? Explain.

6.6 Explain why the Fermi level of an extrinsic semiconductor rises as
the donor concentration is increased, and falls as the acceptor concentration
is increased.

6.7   Explain why the Fermi level in an n-type semiconductor falls with increasing temperature, approaching the middle of the energy gap.

6.8   In a p-type semiconductor, the Fermi level rises with increasing temperature and approaches the middle of the energy gap.   Explain.

6.9   (a) What would you expect the effect of radiation to be on the operation of junction rectifiers and transistors?

(b) How about temperature?

(c) How would you construct devices capable of coping with such environments, if cooling and radiation shielding are not possible?

6.10   Explain how a p-n junction rectifies the flow of holes.

6.11   Draw the energy band diagram for an n-p-n transistor, and explain how it works.

6.12   Consider the junction between an n-type semiconductor and a metal.   Draw the energy band diagram.   If electrons flowed from the semiconductor to the metal, could the device be used to absorb heat, i.e., as a refrigeration element?   Explain.

6.13   What is a Zener diode?   Use the library.   For a start, see p. 279 of Azaroff and Brophy's *Electronic Processes in Materials,* in the Bibliography at the end of this chapter.   How does the Zener diode work and what applications are possible?

6.14   Show that the carrier concentration in an intrinsic semiconductor is proportional to $e^{-\Delta E_0/2kT}$.

6.15   Semiconductors having very high donor or acceptor concentrations, such as those discussed in Section 6.8 for the Esaki diode, have correspondingly high carrier concentrations but, unexpectedly, not very high conductives.   Why?   (*Hint.*   See the sections on mobility and conductivity in Chapter 5.)

6.16   Explain the use of a transistor as a *current* amplifier.

6.17   Design a semiconductor detector for (a) visible radiation, and (b) nuclear radiation.   What physical differences should there be?   What sort of circuits would you use?

CHAPTER SEVEN

Semiconductor Materials

The Group IV semiconducting elements all have the diamond cubic structure. The Group IV, III-V, and II-VI semiconducting compounds generally have the sphalerite or wurtzite structures, which are closely related to the diamond cubic structure. In compounds, carriers may exist because of variable valence or, in the case of transition metal compounds, open states in the d band. Very high purity and perfection are necessary in useful semiconductors. To attain the necessary purity, zone refining is used. Subsequently, single crystals of high perfection are grown, usually by the Czochralski or Bridgeman-Stockbarger methods. Doping to introduce excess carriers may be performed during or after the growth of the crystal. n-type regions may be changed to p-type and vice-versa because of the compensation effect of the impurities. p-n junctions may be grown or made by alloy, diffusion, rate-growing, melt-back, and other techniques. Microelectronic circuits, including many electronic components are incorporated into a single, tiny piece of material by diffusion techniques, combined with oxidation, photolithography, etching, evaporation, and epitaxial deposition.

7.1 INTRODUCTION

The most widely used semiconductors are the two elements lying in Group IV of the periodic table, Si and Ge. Both have the diamond cubic structure. The coordination is tetrahedral, and bonding is essentially covalent. The size of the energy gap increases with the strength of bonding, as shown in the second column of Table 7.1. Group IV elements, carbon (diamond) and

Table 7.1   Some Room-Temperature Properties of Group IV Semiconductors

ELEMENT	E_g (eV) ENERGY GAP	σ_e CONDUCTIVITY ohm^{-1} m^{-1}	μ_e, m^2/V-SEC ELECTRON MOBILITY	μ_h, m^2/V-SEC HOLE MOBILITY
Diamond	5.3	10^{-12}	0.18	0.12
Silicon	1.1	5×10^{-4}	0.14	0.048
Germanium	0.72	2.2	0.39	0.19
Gray tin	0.08	10^6	0.20	0.10

gray tin are included.   Diamond has the strongest bonding and the largest energy gap.   It is an insulator at room temperature.   If diamond is heated to 1000°C, enough carriers can be excited across the energy gap to make it an intrinsic semiconductor.   The other materials, if pure, are intrinsic at room temperature.   Germanium can be readily refined to make it intrinsic.   For silicon this is hardly possible because an impurity level four orders lower than that of germanium is required.   Extrinsic silicon nevertheless is widely used in semiconductor devices.

7.2   SEMICONDUCTING COMPOUNDS

Some covalent Group IV compounds are semiconductors, others insulators; silicon carbide is both.   Above 500°C, SiC is an intrinsic semiconductor when pure.   If small amounts of aluminum are added it becomes *p*-type.   Dissolved nitrogen can make it *n*-type.   The two allotropes of SiC are structurally related to the diamond cubic structure.   A closer similarity to diamond cubic structure is found in compounds of elements from Group III and Group V of the periodic table.   Table 7.2 lists energy gaps and carrier mobilities for such semiconducting III-V compounds.   All of these compounds have the sphalerite (cubic ZnS) structure (discussed in Chapter 3, Volume I), which is closely related to the diamond cubic, as indicated in Figure 7.1*b*.

For AlP, GaAs, and InSb, the structural comparison can be extended as shown in Table 7.3.

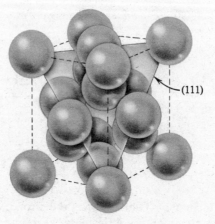

(a) Diamond cubic structure

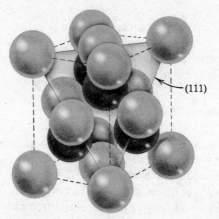

(b) Sphalerite structure. The anion lattice is FCC, with anions in all the upright tetrahedra.

Figure 7.1   Comparison of the diamond cubic and sphalerite structures.

Table 7.3 also shows that the compounds have larger energy gaps than the elements. As mentioned in Chapter 6, a large energy gap is useful because intrinsic behavior, which renders semiconducting devices useless, is displaced to higher temperatures by larger gaps.

Table 7.2   Properties of III-V Compounds

MATERIAL	ENERGY GAP E_g (eV)	ELECTRON MOBILITY m^2/V-SEC	HOLE MOBILITY m^2/V-SEC
AlP	3.0	—	—
AlAs	2.3	—	—
GaP	2.25	0.045	0.002
AlSb	1.52	0.140	0.020
GaAs	1.34	0.85	0.45
InP	1.27	0.60	0.016
GaSb	0.70	0.50	0.085
InAs	0.33	2.30	0.010
InSb	0.18	8.00	0.070

Germanium becomes intrinsic at 100°C; silicon, with a larger energy gap, at 200°C. Some of the compounds of Table 7.2 which have larger energy gaps become intrinsic only at higher temperatures.

It is always desirable to operate semiconducting devices in the exhaustion range where the carrier concentration is relatively insensitive to temperature. Operation is possible in the extrinsic range, yet small temperature changes which drastically change carrier concentrations lead to unstable operation. A large energy gap effectively extends the exhaustion range to higher temperatures. Impurities of very low ionization energy can increase the energy gap. In general, a low impurity ionization energy and a large energy gap are required if the exhaustion range is to be as wide as possible. The III-V compounds listed in Tables 7.2 and 7.3 provide such large energy gaps. In addition, there is more

Table 7.3   Structural Comparison of III-V Compounds and Elemental Semiconductors

	Si	AlP	Ge	GaAs	Sn (gray)	InSb
Unit cell edge	5.42	5.42	5.62	5.63	6.46	6.48 Å
Interatomic separation	2.42	2.34	2.44	2.44	2.80	2.80 Å
Energy gap	1.1	3.0	0.72	1.34	0.08	0.11 eV

flexibility in the kind of donor or acceptor impurity which can be added to type III-V compounds. Si or Ge may be doped with B, Al, Ga, and In (all Group III elements) to make p-type material. P, As, and Sb are added to make n-type material. Both p-type and n-type GaAs may be made by adding Ge. If the Ge atoms substitute for Ga atoms, we have Group IV atoms on Group III sites, extra electrons on the extra bonds, and donor levels. On the other hand, if the Ge atoms sit on As sites, we have Group IV atoms on Group V sites, and acceptor levels. Special heat treatments can make Ge atoms deposit preferentially on either site to give p or n-type GaAs.

It is well to note the very high mobilities for conducting electrons in InSb and InAs, and the relatively high mobilities for both types of carrier in GaAs. High mobility is a useful property, since it leads to high electrical conductivity.

7.3   OTHER SEMICONDUCTING COMPOUNDS

Type II-VI compounds may be insulators as well as semiconductors. Three well-known examples are CdS, CdSe and CdTe with energy gaps of 2.45 eV, 1.8 eV and 1.45 eV, respectively. CdS and CdSe have the wurtzite (ZnS) structure and CdTe the sphalerite structure. The two structures only differ in the stacking arrangement of anions. Figure 7.1b shows the FCC stacking in sphalerite and Figure 7.2 the HCP structure of wurtzite. In both cases all the cations occupy either upward or downward coordination tetrahedra. The energy gap of CdS corresponds to frequencies which lie in the visible region of the spectrum. It is useful in light meters because of its high majority-carrier lifetimes.

Lead sulfide, selenide, and telluride, all IV-VI compounds are also useful in radiation detection. Since their energy gaps are 0.37, 0.27, and 0.33 eV, respectively, they are most useful in the infrared region of the spectrum. Lead telluride is also useful as a thermoelectric material. Many other semiconducting sulfides, selenides and tellurides have been studied. Some are true stoichiometric ternary compounds with specific crystal structures. Others are more like solid solutions. In the latter case the energy gap varies with concentration.

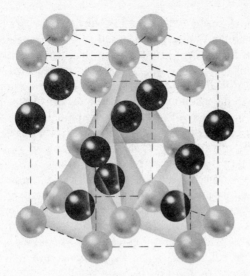

Figure 7.2 The wurtzite structure. This is really the sphalerite structure with the anion sites in an HCP lattice rather than the FCC of sphalerite. Once again, all upright tetrahedra are occupied.

Many oxides are semiconductors. Zinc oxide, with an energy gap of 3.3 eV, is a good example. If the oxide is heated in zinc vapor, zinc atoms dissolve interstitially in the oxide lattice. The zinc atoms ionize readily, and donate electrons to the conduction band. This makes the material an n-type semiconductor. The p-type analog to the zinc oxide case is cuprous oxide, Cu_2O. If a copper ion is *removed* from the lattice, it takes an electron along from the valence band to preserve neutrality. The unsaturated oxygen bond then acts as an acceptor level.

Transition metal oxides are more complicated due to the unfilled d shells of their metal ions. Such oxides should in general be insulators for the oxygen $2p$ shell (in the solid oxides, the $2p$ band) is filled, and the outer levels of the cations are empty due to the transfer of the electrons. The corresponding band is therefore empty. The $3d$ or $4d$ levels may, however, be broadened into bands which are only partially filled. Motion of electrons in the d bands therefore makes conduction possible. This is the reason that the oxide TiO has metal-like properties. If the d electrons participate in bonding, as in TiO_2, two $3d$ electrons are no longer in the $3d$ band, but in the $2p$ band of the extra oxygen. The bonding is then ionic. Since the $3d$ band of pure TiO_2 is empty, this oxide is an insulator. Deviations from ideal stoichiometry can produce holes in the oxygen $2p$ band or conduction electrons in the titanium $3d$ band to make TiO_2 an extrinsic semiconductor. Thus, we can have semiconduction due to transitions to and from the d band which may or may not participate in the bonding. The mobility of carriers in the d band is usually very low; it is of the order of $10^{-6}m^2$/volt-sec. Holes or electrons present in another band may then have a dominating influence.

Semiconduction has also been observed in aromatic hydrocarbons such as anthracene ($C_6H_4:Cl:C_6H_2$) and in some long-chain polymers. In these cases it is necessary that the component molecules interact so that local levels are broadened into bands. The energy gaps of organic semiconductors may be as large as a few electron volts. In some cases the conductivity is nearly metallic and the mobilities are quite small.

7.4   ZONE REFINING

Although it is necessary to have acceptor or donor impurities, a practical semiconductor must, otherwise, be as chemically pure and as crystallographically perfect as possible. Thus, as mentioned in Chapter 6, the p-n junction should occur in a single crystal because a grain boundary at the junction might contribute extra levels and otherwise complicate matters. Recombination centers, traps, and local distortion of the band structure, may all exist around a grain boundary. If a p-n junction occurs at a grain boundary, rectification is hindered or prevented. Moderate or large dislocation densities tend to have similar effects. In transistor materials, the lifetime of excess minority carriers should be as large as possible because the operation of transistors depends on minority carrier injection across the base from the emitter-base junction to the base-collector junction. If the excess minority carriers are arrested in their passage through the base by traps or recombination centers, transistor operation is impossible. Impurities and imperfections are, therefore, highly undesirable. For practical purposes, it is also advantageous to control the resistivity of the material. As Problem 7.1 shows, the typical

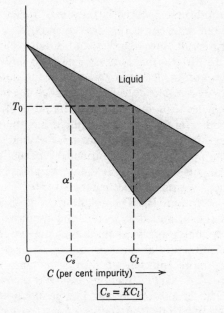

$$\boxed{C_s = KC_l}$$

Figure 7.3 The straight liquidus and solidus imply that the ratio of impurity concentration in the solid to that of the liquid is a constant (K), called the segregation constant.

resistivity of 10^{-2} ohm-m for n-type material implies a donor density of much less than 1 part per million. The level of undesirable impurities must, therefore, be of the order of parts per billion before the donor is added, if real control of the resistivity is to be achieved.

To attain the highest purity possible, zone-refining is employed following orthodox chemical processing. Without zone-refining very little transistor grade material would be available. The zone-refining process is based on the segregation of dissolved impurities during *nonequilibrium solidification.** Figure 7.3 describes the zone-refining process in nearly pure material where only one impurity is present. Let us consider the case where the impurity depresses the melting point as indicated in Figure 7.3. (Problem

* Segregation during normal freezing and zone-refining was analyzed in Section 7.2 and Problems 7.1 to 7.3 of Volume II, and indirectly in Section 8.2 of Volume I. Problem 7.3 of Volume II deals directly with zone-refining.

7.2 deals with the case where the impurity raises the melting point.) Since the concentrations of the impurity in solid and liquid are very small, the liquidus and solidus curves can be approximated by two straight lines. The ratio between the impurity concentration in the solid and that in the liquid is then a constant; $C_s/C_l = K$. The constant K is called the *distribution or segregation coefficient*. If the solidifying liquid is of composition C_l, the first solid to emerge from the liquid has the composition KC_l. If K is less than one, as in Figure 7.3, the solid is purer than the liquid. The material can be purified by partially solidifying and then pouring off the liquid. If the entire piece of material is rapidly solidified, coring will occur (see Chapter 7 of Volume II and Chapter 8 of Volume I). The first material to solidify is again the purest. A rod solidified from one end by moving the solid-liquid interface continuously along leaves the purest and most useful material at the initial end.

A schematic representation of a zone-refining apparatus is shown in Figure 7.4. Only a section of the bar of semiconductor material is melted at any one time. For germanium and silicon and most other materials, the source of heat is usually a high frequency induction coil. As the coil is moved along the bar, the molten zone moves with it; the liquid, which no longer lies in the hot region, solidifies. The length of the zone remains constant if the power input of the induction coil and the environment of the bar remain constant.

In discussing zone-refining, it is convenient to make the following simplifying assumptions (see Chapter 7 of Volume II): (1) due to the rate of solidification, there is no diffusion in the solid, and

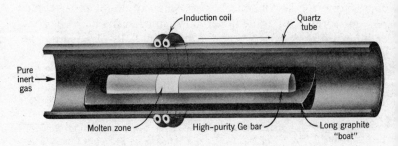

Figure 7.4 Schematic representation of a zone-refining apparatus. The molten zone moves down the bar.

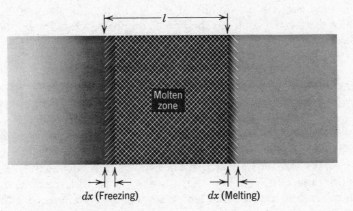

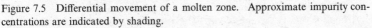

dx (Freezing) dx (Melting)

Figure 7.5   Differential movement of a molten zone.   Approximate impurity concentrations are indicated by shading.

(2) diffusion in the liquid is so rapid that the liquid is perfectly homogeneous.   To refine a hypothetical bar of impurity concentration, C_0 let us assume that melting starts at the left end and the zone is moved to the right, as in Figure 7.4.   Suppose the molten zone of unit cross section moves a distance dx, as shown in Figure 7.5.   At the left end, the material in the volume dx is solidifying. If the concentration of the impurity in the liquid in the zone is C_l, the concentration in the solidifying material is KC_l.   The amount of impurity leaving the zone is $KC_l dx$ at the solidifying (left) end. At the right-hand end of the zone, the section dx is melted.   Since its impurity concentration is C_0, an amount $C_0 dx$ of impurity enters the zone.   Because the rate of accumulation of impurity in the zone is just the input minus the output of impurity

$$l dC_l = (C_0 - KC_l)dx \qquad (7.1)$$

since l is the volume as well as the length of the zone in this case, and $l dC_l$ is just the change in total impurity content of the zone. In terms of the concentration in the solidified material, $C_s = KC_l$:

$$\frac{l}{K} dC_s = (C_0 - C_s)\, dx \qquad (7.2)$$

In integral form this becomes

$$\frac{l}{K} \int_{KC_0}^{C_s} \frac{dC_s}{C_0 - C_s} = \int_0^x dx \qquad (7.3)$$

The first material to solidify at $x = 0$ has the composition KC_0, since the molten zone has the initial composition C_0. Integration then gives

$$C_s = C_0[1 - (1 - K)e^{-Kx/l}] \qquad (7.4)$$

Equation 7.4 is shown graphically in Figure 7.6. Further purification results from repeated melting indicated in Figure 7.7. The impurity level in the left-hand third of the bar can be reduced to a millionth of its previous magnitude by ten passes of the zone: With proper equipment it is not necessary to repeat the run ten times; instead, ten narrow zones in a row, each following the other, are simultaneously melted. Thus, a single pass with ten induction coils has the same effect as ten passes with one coil, provided that the bar is long enough.

A crucible is not necessary if the *floating zone* technique is used. In this case, the bar is vertical and the liquid zone is held in place by surface tension. The possibility of contamination by the crucible is eliminated by this technique (if it is carried out in vacuum or inert gas). The floating zone method is used to refine silicon which has a relatively high melting point and readily reacts with crucible material.

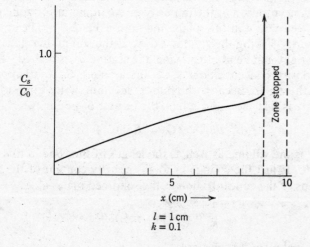

$l = 1$ cm
$k = 0.1$

Figure 7.6   Single-pass zone refining of a 10-cm bar, using Eq. 7.4. The final solidification of the zone is responsible for the peak on the right.

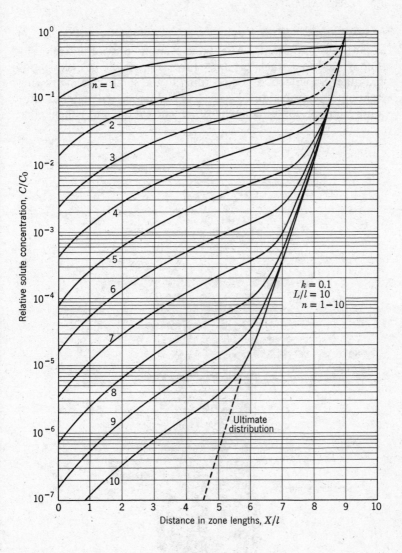

Figure 7.7   The result of repeated zone-refining passes.   (After W. G. Pfann, *Zone Melting*, Wiley, New York, 1958, p. 211.)

7.5   CRYSTAL GROWTH

As mentioned in the previous section, crystal perfection, as well as purity, is important.   After zone-refining, the purified material is converted to a single crystal, as free as possible of dislocations and other imperfections.   According to the single-crystal growing techniques described in Chapter 6 of Volume II the material must solidify under conditions where the nucleation rate is negligible

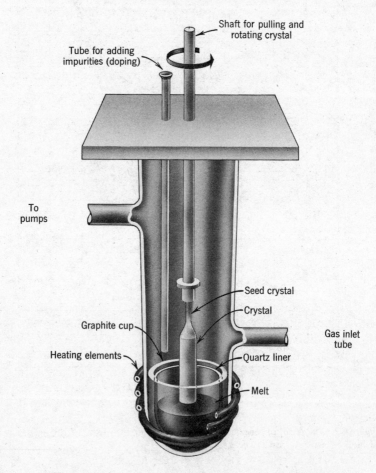

Figure 7.8   Schematic of a Czochralski crystal-pulling furnace.

and the growth rate high. A single nucleus will then grow until the entire material is transformed to a single crystal. The single nucleus can be provided by using a *seed crystal* to initiate growth. Continuous single crystal growth therefore requires (1) very little supercooling and (2) absence of heterogeneous nucleants. The widely used *Czochralski method,* shown in Figure 7.8 answers these requirements. Here the liquid is kept just above the melting point. A seed crystal held just below its melting point, is brought into contact with the liquid. The temperatures of seed and melt are adjusted to make the seed grow slowly (typically, 0.1 to 4 cm per hour). The seed is then withdrawn fast enough to keep the liquid-solid interface near the melt surface. The slower the growth rate, the more perfect the crystal. Crystals of germanium or silicon up to 25 cm can be grown in this manner although 3 cm is sufficient for most applications. The advantage of the Czochralski method is effective control. Because of the geometry of the solid and liquid, temperatures can be precisely controlled. The composition of the liquid can also readily be adjusted. If the composition of the liquid is held constant, the crystal will be homogeneous, even if it has a different composition than the liquid. The composition of the melt must be carefully controlled when making crystal compounds. In the case of PbTe, for example, exact stoichiometry is difficult to achieve. Excess Pb usually turns up. In growing a GaAs crystal, the As vapor pressure over the melt must be controlled. For industrial purposes, Czochralski crystal-pulling devices are usually automated on a large scale.

Another crystal growing method employs the Bridgman-Stockbarger technique illustrated in Figure 7.9. Here a homogeneous hot zone is maintained just above the melting point of the material. As the sharp point of the crucible is slowly lowered out of the hot zone a seed forms at the very tip which is coolest. The small size of the tip leads to the formation of only one nucleus. Thus as the crucible is slowly lowered (usually 0.1-1 cm/hour), the seed grows.

Crystals may be grown from the vapor as well as the liquid. The vapor may be composed of metal atoms, or a volatile compound (usually a halide) of the semiconductor metal which is deposited or decomposed on a seed crystal. Compound semiconductors may be made by mixing the vapors of the two constituents and regulating the partial pressures. CdS and CdSe

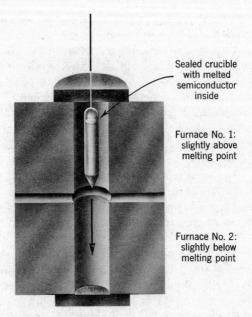

Sealed crucible
with melted
semiconductor
inside

Furnace No. 1:
slightly above
melting point

Furnace No. 2:
slightly below
melting point

Figure 7.9   Schematic of Bridgman-Stockbarger crystal-growing apparatus.

crystals are grown in this manner. The procedure is similar in principle to growth from the melt. Either a seed crystal is used, or a seed is formed by momentarily supersaturating the vapor to a high enough degree. Once the seed is formed, supersaturation is kept low to avoid the formation of other nuclei as growth of the seed proceeds.

Semiconductor single crystals can also be grown by the zone-refining techniques described in the previous section. It is limited in most cases to crystals of small diameter. Since refining can also occur during growth of compounds or solid solution single crystals, special precautions for composition control are imperative.

7.6   JUNCTIONS

Donor or acceptor impurities may be added to the liquid from which the semiconductor single crystal is grown, or they may be deposited later on the crystal surface and diffused inward. In both techniques the intentional addition of impurities is called *doping*. A *p*-type material can be doped with enough donor impurities to convert it to *n*-type, because donor and acceptor levels then neutralize, or *compensate,* each other.

If donors are added to *p*-type material, most of the donor electrons will descend to the empty states (holes) in the valence band. If enough donors are added, most of the holes will be canceled, and conduction electrons will become the majority carrier. The reverse occurs when acceptor impurities are added to *n*-type material. Conduction electrons then descend to the new empty levels and are thus taken out of circulation; holes now become the majority carriers. We can regard compensation as a shifting of the Fermi level. Addition of acceptors lowers the Fermi level, while donors raise the Fermi level. If the Fermi level is above the middle of the energy gap, the material is *n*-type; below it is *p*-type. In general the filling of all states in a compensated semiconductor follows the Fermi-Dirac distribution.

Compensation is essential in the manufacture of junction devices, as shown in Figure 7.10. The earliest technique used is shown in Figure 7.10*a*. It is called the *grown junction* or *double doping* method. As the crystal is grown, the composition of the melt is changed by adding donor and acceptor impurities alternately. In this manner alternating *p*- and *n*-type regions are formed in the crystal.

The simpler *alloy technique* is illustrated in Figure 7.10*b*. A thin slab of *n*-type material such as Ge doped with Sb is used as starting material. Two small In acceptor dots are placed on the surfaces and melted. Near the dots, the germanium dissolves in the In. On cooling, very dilute solid solutions of In in Ge solidify first on the slab. If cooling is very slow, there is no nucleation, the slab serves as a seed, and remains a single crystal. Two *p* regions are produced in this manner to form a *p-n-p* transistor. The dots are on opposite sides of the slab to make the base region thin, since a thin base region makes a more efficient transistor. This is possible because the holes injected across the base from the emitter to the collector travel a shorter distance. This results in less loss by recombination and better high-frequency performance. The base, may also be thinned by etching before processing to give a so-called *microalloy structure.*

Extremely small transistors are made by the *diffusion* technique illustrated in Figure 7.10*c*. There are several steps employed, all

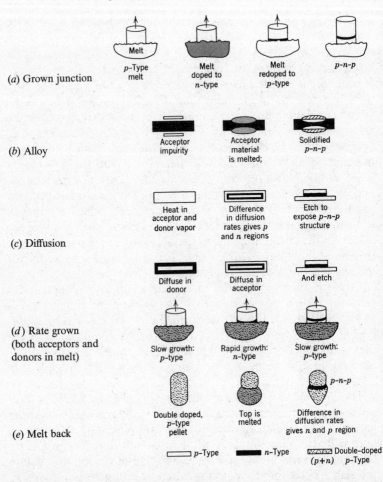

(a) Grown junction

Melt

p-Type melt

Melt doped to *n*-type

Melt redoped to *p*-type

p-n-p

(b) Alloy

Acceptor impurity

Acceptor material is melted;

Solidified *p-n-p*

(c) Diffusion

Heat in acceptor and donor vapor

Difference in diffusion rates gives *p* and *n* regions

Etch to expose *p-n-p* structure

Diffuse in donor

Diffuse in acceptor

And etch

(d) Rate grown (both acceptors and donors in melt)

Slow growth: *p*-type

Rapid growth: *n*-type

Slow growth: *p*-type

Double doped, *p*-type pellet

Top is melted

Difference in diffusion rates gives *n* and *p* region

p-n-p

(e) Melt back

☐ *p*-Type   ■ *n*-Type   ▨ Double-doped (*p+n*) *p*-Type

Figure 7.10   Schematic of transistor-manufacturing techniques. (Courtesy of General Electric Co.)

involving the diffusion of impurities inward from the gaseous phase. Masks are often used to limit the region of deposition. In some cases it is possible to use only one diffusion step because donors and acceptor do not diffuse in at the same rate. In germanium for example, donors diffuse faster than acceptors. In silicon, the reverse is true. On diffusion annealing a *p*-type germanium wafer in a mixed donor-acceptor vapor, the donors penetrate more deeply but spread more thinly than the acceptors. This provides an *n* region where the donors have diffused but the acceptors have not. Since the acceptor density is high near the surface it becomes a *p* region. It is also possible to anneal *n*-type silicon in the same donor-acceptor vapor to form an *n-p-n* silicon transistor.

The difference in the diffusion rates of acceptor and donor atoms makes various other techniques possible. One, in Figure 7.10d, has a melt with a large concentration of the slow-diffusing impurity and a small concentration of the fast-diffusing impurity. In germanium, this signifies a low-donor and a high-acceptor concentration. To grow *n*-type crystals, the growth rate is speeded up and, for *p*-type, slowed down. In the *melt-back process* indicated in Figure 7.10e, the difference in diffusion rate is applied by simply melting the tip of a pellet which contains both donors and acceptors. The resolidified structure then has a *p* and an *n* layer.

The importance of donor or acceptor atom concentration makes precise control of the doping operation mandatory. Such control makes it possible to obtain maximum injection efficiency of the emitter-base junction in transistors discussed in Section 6.7. This is accomplished by doping the emitter as heavily as possible and the base as lightly as possible. The density of majority carriers in the emitter will then be much higher than in the base, and the majority carriers in the emitter will dominate the forward current of the emitter-base junction. Other factors, such as the resistance of the device, its operating voltage, and the frequency at which it should perform, all demand control of the doping operation. In regard to high-frequency performance, the shape of the junction is also important, since it determines the carrier density as a function of distance.

7.7 INTEGRATED CIRCUITS

The small physical size and minimal power requirements of semiconductor devices make it possible to achieve the miniaturization desired in computers and aeronautical as well as astronautical applications. With the development of crystal growing, micromachining, etching, vacuum deposition, and diffusion techniques, it has become possible to integrate components and circuits into single functional units. Whole amplifier, communication, and computer sub-assemblies can now be built from single semiconductor wafers utilizing semiconductor principles and techniques. The unit shown in Figure 7.11 has an area of 0.2 cm² yet contains the equivalent of 31 conventional transistors, diodes, resistors, and capacitors.

In the fabrication of integrated circuits, such components as resistors, capacitors and connections are made by vacuum deposition or sputtering of metal or insulator films on tiny glass or ceramic plates. The resistors are usually made of sputtered tantalum or evaporated zirconium-nickel alloy films. The dielectric of the capacitor elements is either tantalum oxide (Ta_2O_5) or silicon monoxide (SiO). Silicon dioxide (SiO_2) films are used for insulation and vacuum deposited aluminum or other short and narrow metal film strips as connections. Masks and stencils are commonly used to limit film deposits, and photolithography techniques make precise chemical etching possible. Figure 7.12a indicates the conversion of a single transistor element into two transistors connected together at the collector (or emitter). Figure 7.12b illustrates the fabrication of a simple circuit containing a diode transistor and resistor. The n, p, and n layers are diffused into the intrinsic crystal in succession. The impurity elements are deposited on the surface through masks or

Figure 7.11   A micro-circuit, ⅛ in. × ¼ in. over-all, with 31 components (transistors, capacitors, and resistors) represented in it. (Courtesy Texas Instruments, Inc.)

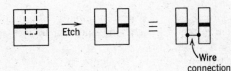

(*a*) The etched transistor is equivalent to two with the collectors connected, but the wire connection is now unnecessary.

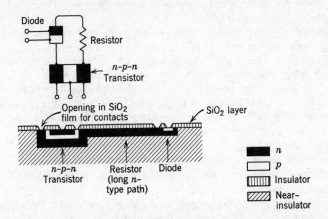

(*b*) The conventional circuit above is actually much larger than the circuit below, which was created by successive diffusions (with masking) into intrinsic germanium or nearly intrinsic silicon; called a "near-insulator."

Figure 7.12   Simple examples of the functional electronics approach.

stencils. Three successive evaporation and diffusion steps are required in this case. Bare spots are left in the SiO_2 insulating film for connections. Narrow films of aluminum evaporated on the SiO_2 layer serve as connectors.

For detailed discussion of integrated circuits and semiconductor technology, the reader is advised to consult the more specialized references listed below.

DEFINITIONS

Zone Refining.   The segregation of impurities to one end of a bar by the passage of a molten zone.

Distribution Coefficient (K).   For a given impurity, the ratio between the impurity concentration in the solid and that in the liquid in equilibrium; $K = C_s/C_l$

Doping.   The intentional addition of donor or acceptor impurities to a semiconductor.

Seed Crystal.   A piece of single crystal which is used as a nucleus to grow a larger crystal.

Czochralski (Crystal Pulling) Technique.   The withdrawal of a crystal from the melt as fast as it grows.

Compensation.   The neutralization of donor impurities by acceptor impurities and vice versa.

Grown Junction Transistor.   The growth of an *n-p-n* or *p-n-p* structure by successive additions of donors and acceptors to the melt.

Alloy Transistor.   The emitter and collector regions are formed by solidification of melted "dots" of the base material dissolved in the dopant.

Microalloy Transistor.   An alloy transistor with a base which has been thinned, usually by etching, before processing.

Diffused Junction Transistor.   The collector and emitter regions are formed by diffusion of additives diposited from the vapor phase.

Rate Grown Transistor.   *p* and *n* regions are created by varying the growth rate of the crystal.

Melt-back Transistor.   Two of the *p* or *n* regions form during the solidification of the melted tip of a semiconductor pellet.

BIBLIOGRAPHY

SUPPLEMENTARY READING:

Azaroff, L. V., and Brophy, J. J., *Electronic Processes in Materials,* McGraw-Hill, New York, 1963, Chapter 9, pp. 291–294 and 300–302.

Biondi, F. J., *Transistor Technology,* I-III, Van Nostrand, New York, 1958.

General Electric Transistor Manual, Fifth Edition, 1960, General Electric Company, Chapter 2.

Gilman, J. J., Ed., *Art and Science of Growing Crystals,* Wiley, New York, 1963, pp. 343–398.

Kingery, W. D., *Introduction to Ceramics,* Wiley, New York, 1960, pp. 665–685.

Keonjian, E., *Microelectronics,* McGraw-Hill, New York, 1963.

Van der Ziel, A., *Solid State Physical Electronics,* Prentice Hall, New York, 1957, pp. 269–282.

PROBLEMS

7.1   What donor concentration in *n*-type germainium will result in a resistivity of 10^{-2} ohm-meter?   Use the data of Table 7.1.   Assume that one half of the donors are ionized.   What concentration of acceptors in *p*-type

Ge is necessary for the same resistivity? Give your answer in parts per million.

7.2   Describe the effect of zone-refining on the distribution of an impurity for which K is more than one. Draw typical zone-refining curves.

7.3   Use the answers to Problem 7.2 in this problem. Niobium is zone-refined, and the resistivity ratio measured as a function of distance along the bar. It is found that the ratio rises from the starting end to the final end. After referring to an atlas of phase diagrams (such as *The Constitution of Binary Alloys* by M. Hansen and K. Anderko, McGraw-Hill, New York, 1958), it is concluded that tungsten, tantalum, or molybdenum are being refined out. Why? What does the variation of the resistivity ratio say about K, and what does K say about the Nb-rich end of the pertinent binary phase diagram? (It is not necessary to refer to Hansen and Anderko to answer this problem, but it might prove profitable.)

7.4   Show that the distribution coefficient is constant if the liquidus and solidus are both straight.

7.5   Describe briefly the steps used in chemical refining of (a) Ge, and (b) Si preceding zone refining.

7.6   What features would you want in an automatic Czochralski apparatus? What variables should be controlled, why, and how would you try to do it?

7.7   Do the same for an automatic zone-refining unit.

7.8   The injected minority carriers move across the base region of a transistor by diffusion. The fraction of carriers that reaches the collector falls off with the frequency of the signal applied to the emitter-base junction. Some arbitrary fraction, say $\frac{1}{2}$, is set as the minimum fraction tolerable, and the frequency corresponding to this fraction is called the "cut-off frequency." The cut-off frequency varies approximately as the inverse square of the base thickness. Use diffusion theory to explain this. Refer to pages 79 and 80 of Volume II. If the diffusion coefficient is proportional to the mobility, how does the cut-off frequency depend on minority carrier mobility in the base?

7.9   Using the results of Problem 7.8, how would cut-off frequencies of n-p-n and p-n-p transistors of Ge compare? Si? InSb?

7.10   Do the semiconducting elements and compounds obey the 8-N rule? Review Chapters 1 to 3 of Volume I.

7.11   An intrinsic semiconductor is doped with donors. Then, just enough acceptors are added to compensate the donors. How does the Fermi level change? Draw band diagrams including the Fermi level for the semiconductor after each doping. Are there any differences between a compensated and an intrinsic semiconductor?

7.12   Explain the mechanism of compensation that occurs when a donor majority exists originally but is compensated by the addition of acceptors.

7.13   Why do we desire a large energy gap between the valence and conduction bands and a small impurity ionization energy in semiconductor materials to be used in practical devices?

7.14   Consider an exactly compensated semiconductor in which the hole and conduction electron densities are equal, and the Fermi level lies in the middle of the energy gap. Is this semiconductor intrinsic? Explain your answer.

7.15   To what frequencies or wavelength of visible light do the energy gaps in CdS, CdSe, and CdTe correspond?

7.16   Draw a simple circuit illustrating the use of CdS as a sensor in a light meter.

7.17   What advantages and disadvantages would you expect to belong to an organic semiconductor?

7.18   Suppose the concentration of donor atoms was twice that of the acceptor atoms in a semiconductor. What can you say about the carrier concentration? What other data is necessary to a complete answer?

7.19   Suppose that you had all of the necessary data mentioned in the previous problem. Outline and derive an expression for the carrier concentrations.

7.20   Show that, for dilute carrier concentrations, the hole and electron concentrations follow a "mass action" law, i.e.,

$$n_h n_e = \text{const.}$$

in extrinsic semiconductors.

7.21   Describe briefly what is meant by epitaxial growth of a vacuum deposited film.

CHAPTER EIGHT

Thermoelectricity

In a common thermocouple, a voltage is induced between the hot and cold junctions of two dissimilar metals which increases monotonically with increasing temperature difference. The temperature difference produces changes in the distribution of electrons in the energy states at the hot and cold ends, leading to a flow of electricity. Three experimental observations of thermoelectricity, the Seebeck (1822), Thomson, and Peltier (1834) effects, are related thermodynamically by the Kelvin (Thomson) relations. In semiconductors, much larger thermoelectric effects occur than in metals. Thermoelectric temperature measurement, refrigeration, and heating, and generation of electrical power are all in limited use today. The efficiency of a thermoelectric device depends on the Seebeck coefficient, thermal and electrical conductivity, and the operating temperatures.

8.1 INTRODUCTION

Consider the conducting rod shown in Figure 8.1 with one end maintained hot, and the other cold. At the hot end, electrons will, on the average, be excited to higher energies; the Fermi-Dirac distribution will have more electrons above E_F and fewer below. The higher-energy electrons at the hot end are able to lower their energies by diffusing to the cold end. Thus, the cold end becomes negatively charged, the hot end positively charged, and a voltage is induced along the rod. The induced voltage causes an electrical current to flow, which is equal to the voltage divided by electrical resistance of the rod. Equilibrium is reached when the current is equal to the flow of electrons due to the temperature difference, and the induced voltage is constant. It should be

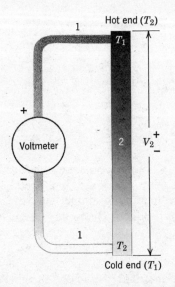

Figure 8.1   Schematic of circuit used to measure thermal electromotive force.

noted that the induced voltage between the ends of the rod depends *only* on the temperatures of the ends. Whatever the distribution of temperatures, the same voltage will appear between the ends.

To measure the induced voltage, a volt-meter is connected to the ends of the rod as shown in Figure 8.1. If the connecting wires, *1*, are of the same material as the rod, *2*, the temperature difference will induce the same voltage in the connecting wires as in the rod, and there will thus be no voltage across the meter. On the other hand, if the connectors are of another material, a different voltage will be induced in *1*, and the net voltage, V_1-V_2, will be observed at the meter. If we raise the temperature difference T_1-T_2 a small amount, ΔT, the induced voltages will rise a small amount. The net voltage, designated as V_{12}, will change. The rate of change of voltage with temperature is defined as the *thermoelectric power, S_{12}* of the junction *1-2:*

$$S_{12} = \frac{dV_{12}}{dT} = \frac{dV_1}{dT} - \frac{dV_2}{dT} \ or \ S_1 - S_2 \qquad (8.1)$$

Although S_{12} is called the thermoelectric power of the junction, it is not at all a property of the junction. It depends only on the bulk properties, S_1 and S_2, of the two materials. The term *power*, though inappropriate, is nevertheless widely used.

The thermally induced voltage, V_{12}, called the *Seebeck potential*, is often used in measuring temperature. One junction of the thermocouple is held at a known temperature. The voltage, V_{12}, is then a function of the temperature of the other junction. Thermoelectric powers have been compiled for a large number of materials and pairs, and calibrated thermocouple wire is a commercial item. In common use are copper (up to 315°C), Constantan (60 Cu-40 Ni) and iron (both up to 950°C), Chromel (90 Ni-10 Cr) and Alumel (94 Ni-2 Al-3 Mn-1 Si), both up to about 1200°C. For higher temperatures up to 1500°C, couples of platinum and various platinum-rhodium alloys are used, and above 1500°C, tungsten-rhenium alloys.

At room temperature and above, thermoelectric potentials in the millivolt range are common; at low temperatures, the potentials are normally in the microvolt range. The thermoelectric power is very sensitive to structural defects so that thermocouple materials must be very carefully processed. In general, anything which increases the resistivity, such as cold work, will increase the thermoelectric power.

In view of the way that the voltages are defined in Figure 8.1, and in the first paragraph of this section, it is expected that the thermoelectric power of a conductor or of any material in which conduction electrons are the majority carrier will be negative. Since only differences in thermoelectric potential are measurable, it is difficult to determine directly the absolute thermoelectric power. Fortunately, the power is zero for a superconducting material, (superconductivity is discussed in Chapter 11) which provides a standard. The absolute thermoelectric power can also be calculated from the Thomson heat as discussed below.

8.2   THOMSON EFFECT

Referring again to the rod of Figure 8.2 with one end hot and the other cold, the ends will be charged, and there will be a voltage along the length of the rod. If conventional electrical current passes down the rod from the hot end to the cold end, electrons

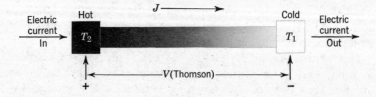

Figure 8.2 Schematic of Thomson effect. In addition to the Ohm's law voltage, a second voltage is observed, due to the temperature difference. When the current is passed, additional heat is absorbed or emitted, depending on the directions of the current and thermal gradient.

flow from cold to hot; as they flow from the negative to the positive end of the rod. This increases their potential energies by absorbing energy in the form of heat. If the current flow is reversed, the electrons will decrease their potential energies by emitting heat. This absorption or evolution of heat must be added to the I^2R power loss or Joule heat together with the heat flow due to the thermal gradient in the rod, to get the total heat production and flow. The absorption or evolution of heat is called the *Thomson effect,* and the relation

$$\frac{dQ(\text{Thomson})}{dt} = -\mu_T J_x \frac{dT}{dx} \qquad (8.2)$$

defines the *Thomson coefficient,* μ_T. J_x and dT/dx are the current density and thermal gradient along the rod of Figure 8.2, and dQ/dt is the heat evolved per unit volume per second. (Note that the Q in this definition has a sign opposite to that common in thermodynamics. Here, it is positive when the heat is emitted.)

8.3   PELTIER EFFECT

A third important thermoelectric effect illustrated in Figure 8.3 is the *Peltier effect.* When current is passed through the thermocouple junction, heat is absorbed or evolved, depending on the direction of the current. The *Peltier heat,* π_{12}, is defined as the reversible heat evolved at the junction per unit time per unit electric current flowing from *1* to *2*, or

$$\frac{dQ(\text{junction})}{dT} = \pi_{12} \qquad (8.3)$$

From this definition, $\pi_{12} = -\pi_{21}$.

In the discussion of the Thomson effect, it was stated that electrons can convert heat to potential energy and vice versa. In general electrons can transport heat in the forms of thermal energy and potential energy. Precisely how much heat the electrons carry depends on temperature, electrical fields, and the material through which the electrons flow. The heat transported per mole of electrons is different for different materials, and therefore their thermoelectric powers and Thomson coefficient differ. When electrons flow from material *1* to material *2* in Figure 8.3, they change the amount of heat that they carry, and the difference must be absorbed or given off, depending on the direction of current flow at the junction. Consequently, the Peltier effect is proportional to the difference in the heat carried by the conduction electrons of the two materials. The absolute Peltier heat may be defined in the same manner as the absolute thermoelectric power

$$\pi_{12} = \pi_1 - \pi_2 \qquad (8.4)$$

Neither the Peltier nor Seebeck effects depends on the *nature* of the junction between the two materials. The junction may be soldered, brazed, spot-welded, or fused. The effects are bulk material and not junction properties.

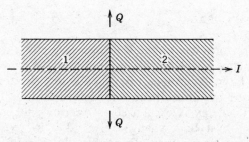

Figure 8.3   Schematic of the Peltier effect.   If current is passed through the 1-2 junction in one direction, heat is evolved.   If the current direction is reversed, heat is absorbed.

8.4 KELVIN (THOMSON) RELATIONS

The three parameters which describe thermoelectric behavior are the thermoelectric power S, the Thomson coefficient μ_T, and the Peltier heat π. These quantities are not independent. Thomson applied the First and Second Laws of Thermodynamics to show that only one of the three is needed to specify the other two. Although Thomson's derivation was not rigorous, since classical thermodynamics is not directly applicable to this problem,* his answers are correct.

$$\mu_T = T \frac{dS}{dT} \tag{8.5}$$

$$\pi = TS \tag{8.6}$$

See Problem 8.16. (Remember that S here refers to the thermoelectric power, not the entropy.) If S is known as a function of temperature, Equations 8.5 and 8.6 give π and μ_T. If μ_T rather than S is known Equation 8.5 can be integrated to obtain S

$$S(T)\text{-}S(0) = S(T) = \int_0^T \frac{\mu_T}{T} \, dT \tag{8.7}$$

$S(0)$ can be dropped from Equation 8.7 because the Third Law of Thermodynamics implies that all thermoelectric effects vanish at $0°\text{K}$. It is then possible, from Equation 8.7 to calculate the absolute thermoelectric power of a conductor at any temperature.

8.5 PHONON DRAG

In the previous description of thermoelectricity it was assumed that the energy of conduction electrons may be converted freely to heat or lattice vibrations. In other words, electrons colliding with phonons give up their energy. The reverse was also assumed to hold. If energy is exchanged in electron-phonon collisions, momentum must be exchanged also. Let us re-examine the thermoelectric experiment of Figure 8.1. Heat will flow from the high-temperature junction (T_2) to the low-temperature junction

---

* The techniques of *irreversible thermodynamics* are necessary. See, for instance, Part III of *Thermodynamics* by H. B. Callen, Wiley, New York, 1960.

(T_1) providing a net phonon flow. Phonon collisions with electrons tend to drive the electrons along as well. Electrons tend to migrate to the cold end from equilibrium considerations alone. The heat flow will also drive an additional number of electrons to the cold end. This effect is called *phonon drag*. Two contributions to the thermoelectric power are therefore observed. The phonon-drag contribution is only considerable at low temperatures. It has a maximum below about $100°$K and falls off rapidly at higher temperatures, and vanishes at $0°$K. Phonon-drag components must also be included in the Peltier and Thomson heats of Equation 8.5 and 8.6. The flow of electrons will *push* heat along the rod, and from junction to junction, by collisions with phonons as well as by the equilibrium effects.

8.6   THERMOELECTRICITY IN SEMICONDUCTORS

In semiconductors, thermoelectric effects are many times larger than in metals. The three coefficients, S, μ_T, and π, may be positive or negative, whereas in the ideal simple conductor, they are negative. Such behavior results from arrangement of the energy bands in semiconductors and the details of the occupation of the states by electrons. To account for the algebraic sign of the three coefficients in semiconductors, let us recall Section 8.1. If electrons are the majority carrier, S is negative. According to Equations 8.5 and 8.6, π and μ_T should, also be negative. If holes are the majority carrier, more electrons are excited into the acceptor levels at the hot end, and thus more holes are available. Electrons at the cold end near the top of the valence band can lower their energies by moving into the holes at the hot end. The hot end becomes negatively charged, and the cold end positively. This behavior is opposite to that discussed in Section 8.1. Therefore, in p-type semiconductors, S, π, and μ_T are all positive. A similar consequence prevails in the Hall effect (Chapter 5). Indeed, calculations show that the Hall coefficient can be determined from the thermoelectric coefficients and vice versa.

The thermoelectric powers of semiconductors are two orders of magnitude larger than for nonferromagnetic metals (refer to Table 8.1). One reason is that the carrier density of semiconduc-

Table 8.1 Thermoelectric Power, Thermal Conductivity, and Electrical Resistivity of Various Metals and Semiconductors at Various Temperatures

MATERIAL	TEMPERATURE ($^{\circ}$C)	S(MICROVOLTS/ DEG K)	ρ (ohm-m)	σ_T(WATTS/ m-DEG K)
Metals (Nonferromagnetic)				
Al	100	-0.20	2.85×10^{-8}	210
Cu	100	$+3.98$	1.7×10^{-8}	390
Ag	100	$+3.68$	1.59×10^{-8}	420
W	100	$+5.0$	0.55×10^{-8}	167
Semiconductors				
$(Bi, Sb_2)Te_3$	25	$+195$	9×10^{-6}	1.47
$Bi_2(Te,Se)_3$	100	-210	10.5×10^{-6}	1.6
ZnSb	200	$+220$	23×10^{-6}	1.6
InSb	500	-130	4.5×10^{-6}	7.5
Ge	700	-210	14.5×10^{-6}	20
TiO_2	727	-200	148×10^{-6}	3.4
$(Li_{0.05}Ni_{0.95}O)$	1100	$+225$	80×10^{-6}	3.5

tors is sensitive to temperature, and the hot end will have more conduction electrons per unit volume than the cold end, or more holes, depending on whether we are dealing with a *p*-type or an *n*-type material. The more important reason arises from the forbidden energy band. In an ideal conductor, most of the carriers are near the Fermi level. In an *n*-type extrinsic semiconductor, the majority carriers occupy states in the conduction band, high above the Fermi level. Consequently the heat transported by an electrical current in an *n*-type semiconductor exceeds the heat transported by the same current in an ideal conductor. The application of this reasoning to the junction between an *n*-type semiconductor and a metal is shown in Figure 8.4. The average energy of the conduction electrons decreases considerably as the junction is crossed and energy is lost in the form of heat. If the flow of current were reversed, the conduction electrons would increase their average energy by absorbing heat.

In the case of *p*-type semiconductors, the carriers lie well below the Fermi level in the valence band, and the heat transported is far less than in the ideal conductor. The large difference, once more, gives rise to a large Peltier effect, but of the opposite sign. A

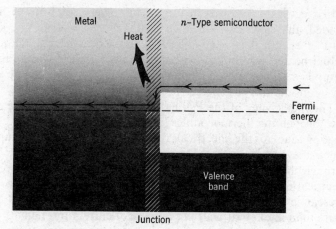

Figure 8.4 The Peltier effect at a junction between a metal and an n-type semi-conductor. The average energy of the conduction electron changes as the junction is crossed. For flow of electrons to the left, heat is given off, equivalent to the reduction in energy. If the flow were to the right, the average energy would increase by the absorption of heat at the junction.

better heat- or cold-producing junction than that of Figure 8.4 is therefore a *p-n* semiconductor junction.

The rapid diffusion of the carriers, together with the high electrical resistivities further increases the thermoelectric effects in semiconductors. Since the thermoelectric voltage is determined by the equilibrium between the diffusion of electrons from hot to cold, and the reverse flow of electrons due to the induced voltage, high diffusivity builds up large voltages, and high resistivity reduces the return flow.

8.7   THERMOELECTRIC HEATING AND REFRIGERATION

By using the Peltier effect, a silent, compact refrigerator or heater may be constructed. As the current passes through the thermocouple, heat is absorbed at the cold junction and emitted at the hot junction; the reverse occurs when the current is reversed. In other words, the thermocouple *pumps* the heat from the cold junction to the hot junction. Peltier heating can be more efficient than electrical-resistance heating. For every kilowatt of heat needed, the electrical-resistance heater must consume one kilowatt of electrical power which is dissipated as heat. The ideal Peltier heater need not do so, since it uses electrical work only to pump heat energy uphill like any refrigerator or heat engine, and uses considerably less power.

The inefficiency of the Peltier heat pump is caused by the heat lost through the thermocouples from the hot to the cold side, and heating due to the electrical resistance of the thermocouples. Since the heat transferred is proportional to the current flow through the thermocouple, large currents are desirable; however, electrical-resistance heating limits the size of the current. At present, even under ideal conditions, it is not possible to pass more than about 0.01 watt per ampere of current through a single thermocouple. Useful devices require a large number of thermocouples connected in series, as shown in Figure 8.5. The elements used are alternating p- and n-type semiconductors to give the maximum difference in π, which is positive and negative, respectively, for p- and n-types. Peltier pumps are not as efficient as conventional mechanical heat pumps.

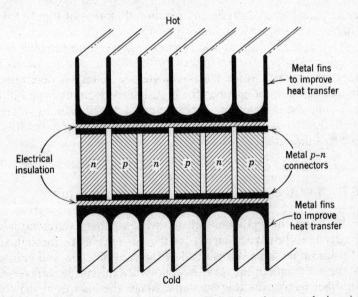

Figure 8.5   A section of a Peltier heat pump, showing three thermocouples in series.

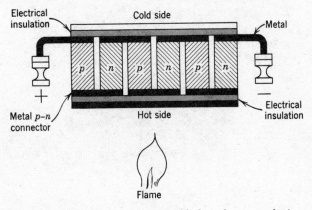

Figure 8.6   A thermoelectric generator, with three thermocouples in series.

8.8   THERMOELECTRIC GENERATORS

The Peltier refrigerator develops a temperature difference by using electrical energy. It is alternatively possible to develop a voltage from a temperature difference by applying the Seebeck effect. The devices look much the same as the Peltier refrigerator of Figure 8.5. For a temperature difference of 100°C, it is possible to develop up to 50 millivolts using certain *p-n* semiconductor combinations. Ordinarily many junctions are used in series (Figure 8.6) and temperature differential of several hundred degrees is required. Although thermoelectric generators are not efficient, they are compact and easily installed and operated. Figure 8.7 shows the use of a thermoelectric converter on the chimney of a kerosene lamp. These generators are reported to be in use in the far northern and agricultural regions of the USSR where electricity is not otherwise available. Similar generators may be operated in semi-arid regions using solar heat.

8.9   THE FIGURE OF MERIT

The ultimate efficiency of a thermoelectric generator or refrigerator is that of the Carnot cycle. However, practical thermoelectric devices are irreversible and dissipate energy by the spontaneous flow of heat through the thermocouple from the hot to the

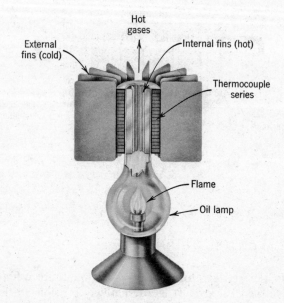

Figure 8.7 Thermoelectric generator powered by an oil lamp. The thermocouples are arranged in a manner similar to Figure 8.6. [After V. Daniel-Bek et al., *Radio* Vol. 2, p. 24 (1954).]

cold junction, and I^2R losses due to the electrical resistance of the thermocouple. The optimum efficiency of a thermoelectric device is therefore the product of the Carnot efficiency and a factor which accounts for heat flow and I^2R losses. For a thermoelectric refrigerator which extracts heat at the lower temperature T_L and emits it at the higher temperature T_H, the optimum efficiency is

$$E_r(\text{optimum}) = \frac{\text{heat absorbed at } T_c}{\text{work done by the current}}$$

$$= \frac{M(T_H + T_L) - 3T_H + T_L}{2(T_H - T_L)(M + 1)} \tag{8.8}$$

where

$$M \equiv (1 + ZT_H)^{1/2}$$

For a thermoelectric generator, the optimum efficiency is

$$E_g(\text{optimum}) = \frac{\text{electrical work done}}{\text{heat absorbed at } T_H}$$

$$= \frac{2(T_H - T_L)(M + 1)}{M(T_H + T_L) - 3T_H + T_L} \qquad (8.9)$$

The parameter Z is called the *figure of merit* for thermoelectric materials. It is defined by the relation

$$Z = \frac{S^2 \sigma_e}{\sigma_T} \qquad (8.10)$$

To increase the efficiency, the figure of merit must be increased. Either the thermoelectric power or the electrical conductivity must increase, or the thermal conductivity must decrease. The latter two will decrease the irreversible thermal and electrical losses. Table 8.2 lists figures of merit for various thermocouples. Where two different materials are involved, averages are taken.

From Chapters 3 and 4 we have seen that for pure metals, σ_e and σ_T are related through the Lorenz factor and temperature by the Wiedemann-Franz ratio. Thus, the possibilities of increasing the efficiencies of metallic thermocouples by increasing Z are very limited. Furthermore, the thermal conductivity of diamond cubic structures is relatively high which makes many of the common semiconductors discussed in Chapter 7 unsuitable for thermoelectric applications.

Bismuth and antimony tellurides or selenides have high figures of merit. In 1956, Ioffe discovered that semiconductor alloys such

Table 8.2   Figures of Merit for Various Thermocouples

MATERIAL	$Z \, (°K)^{-1} \times 10^{-3}$
Chromel-Constantan	0.1
Sb-Bi	0.18
Sb-(91% Bi, 9% Sb)	0.23
ZnSb-Constantan	0.5
PbTe(p-type)-PbTe(n-type)	1.3
$Bi_2Te_3(p)$-$Bi_2Te_3(n)$	2.0
p-n Alloy thermocouples:	
(50% Bi_2Te_3, 50% Sb_2Te_3)	2.5
(75% Bi_2Te_3, 25% Bi_2Se_3, + minor $CuBr_2$ addition)	2.5

as $Bi_2Te_3 + Bi_2Se_3$ has lower thermal conductivities but substantially the same electrical conductivities as the alloy components. Thus, semiconductor alloys offer improvement in the efficiencies of thermoelectric refrigerators and generators.

Other methods have been applied to improve the efficiency. Operation at higher-temperature differences will increase the Carnot efficiency of a thermoelectric generator (Figure 8.8 and Equation 8.9). Different types of semiconductors can be assembled to operate over a range of temperatures. The lower-temperature thermocouples then use the waste heat from the high-temperature thermocouples, and each thermocouple operates in its optimum temperature range.

The optimal efficiency of a thermoelectric refrigerator decreases as the temperature difference increases according to Equation 8.8. Since the efficiency of the refrigerator vanishes when:

$$T_H - T_L = \Delta T(\text{max}) = T_H - T_L \equiv \Delta T(\text{max})$$

$$\Delta T(\text{max}) = 2T_H\left(\frac{M - 1}{M + 1}\right) \quad (8.11)$$

any thermocouple has a $\Delta T(\text{max})$ which it cannot exceed. To

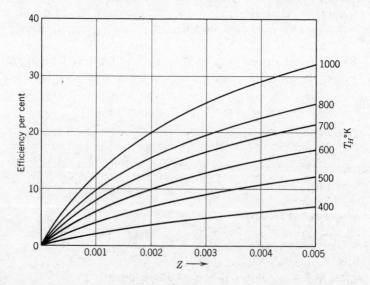

Figure 8.8   Thermoelectric generator efficiency as a function of $Z =$ figure of merit, $T_H =$ hot junction temperature °K, cold junction at 300°K. (Based on data of Ioffe.)

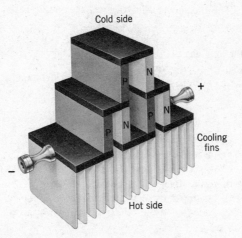

Figure 8.9 Schematic of a cascaded Peltier refrigerator. The two lower stages remove heat generated in the upper stage as well as heat extracted from the cold side. [After B. J. O'Brien et al., *J. App. Phys.*, Vol. 27, p. 820 (1956).]

attain larger temperature differences, the thermocouples are *cascaded,* as shown in Figure 8.9. The use of the two layers illustrated does not allow a doubling of the temperature difference, for the lower layer must remove the heat generated in the upper layer as well as pump the heat from the cold side.

Inspite of intensive scientific and technical development, the highest figure of merit attained thus far is only 3 and the thermoelectric efficiency not greater than 10 per cent. The latter figure is based on the thermal energy absorbed at the hot junctions. In practical thermoelectric generators, the efficiency of heat transfer to the hot junction is itself but 50 per cent which depresses the over-all efficiency to about 6 per cent.

DEFINITIONS

Thermocouple.   Two different conducting materials connected in a looped circuit, with one junction at a higher temperature than the other.
Seebeck Effect.   The generation of voltage by a thermocouple.
Thermoelectric Power(S) (Seebeck on thermal emf coefficient).   The voltage developed in a thermocouple for a unit increase in temperature differences.   The absolute thermoelectric power may be obtained for a single

material. The thermoelectric power of the couple is the difference of the thermoelectric powers of the two materials.

Thomson Effect. The reversible absorption or emission of heat which occurs when an electrical current flows through a temperature difference in a single conductor.

Thomson Heat (μ) (Thomson coefficient). Defined by the relation

$$\frac{dQ}{dt} = -\mu_T J_x \frac{dT}{dx}$$

where Q is the heat evolved per unit volume.

Joule Heat. The electrical resistance loss $I^2 R$.

Peltier Effect. The reversible emission or absorption of heat at the junctions of a thermocouple when a current passes.

Peltier Heat (π). Defined by the relation $dQ/dt = \pi_{12}$ for the junction.

Kelvin (Thomson) Relations. The relations between π, μ_T, and S, $\pi = TS$, $\mu_T = T (dS/dT)$.

Phonon Drag. The electrical current produced by heat flow as a result of phonon-electron collisions.

Figure of Merit(Z). $Z \equiv S^2 (\sigma_e/\sigma_T)$ is used to compare the efficiency of thermoelectric materials.

BIBLIOGRAPHY

Supplementary Reading:

Dike, P.H., *Thermoelectric Thermometry,* Leeds and Northrup, 1954.

Ioffe, A. F., *The Revival of Thermoelectricity, Scientific American,* November 1958, p. 222.

MacDonald, D. K. C., *Thermoelectricity: An Introduction to the Principles,* Wiley, New York, 1962, pp. 1–24.

Goldsmid, H. J., *Applications of Thermoelectricity,* Methuen, Wiley, New York, 1960.

Stanley, J. K., *Electrical and Magnetic Properties of Materials,* American Society for Metals, Metals Park, Ohio, 1964, pp. 167–173.

Weber, R. L., *Heat and Temperature Measurement,* Prentice Hall, 1950, Chapter 5.

Advanced Reading:

Cusak, N., *The Electrical and Magnetic Properties of Solids,* Wiley, New York, 1958, Chapter 5 and also pp. 224–226.

Ioffe, A. F., *Physics of Semiconductors,* Academic Press, New York, 1960.

PROBLEMS

8.1 Design and describe an experimental arrangement for the measurement of the thermoelectric power. There are several techniques which will improve your experiment: (a) a potentiometer, (b) a standard cell, and

(c) a cold reference junction.

For the use and definition of these techniques, see, for instance, *Experimental Metallurgy,* by A. U. Seybolt and J. E. Burke, Wiley, New York, 1953, Chapter 2; also, *Temperature, Its Measurement and Control in Science and Industry,* Reinhold, New York, 1941.

8.2   Design and describe a method for calibrating thermocouples for accurate temperature measurement. Use the same techniques as Problem 8.1: potentiometer, standard cell, and cold junction. The literature mentioned above will also be useful. Consider three cases: (a) copper-Constantan, for use up to 315°C, (b) iron-Constantan, for use up to 980°C, and (c) platinum (platinum-10% Rhodium), for use to 1500°C.

8.3   The Thomson coefficient of lead can be set to zero arbitrarily, and lead used as a standard; thermocouples made of lead and another material will have thermoelectric powers which are, on the *lead standard scale,* just the absolute thermoelectric powers of the other material. Here are some typical data (thermocouple voltages):

Material	Thermal EMF referred to Pb
Al	$-0.47T + 0.003T^2$ microvolts
Cu	$+2.76T + 0.012T^2$ microvolts
Fe	$16.6T - 0.030T^2$ microvolts
Pt	$-1.79T - 0.035T^2$ microvolts

(a) Compute the thermoelectric powers of each, on the lead scale.
(b) Compute the Thomson coefficients.
(c) Compute the Peltier heats.

8.4   Consider a rod with a thermal gradient of 10°C per meter along its axis. One ampere is passed down the rod from the hot end to the cold end and, for one meter of rod, 1.07 milliwatts are emitted. When one ampere is passed in the opposite direction, 0.93 milliwatts are emitted. Calculate the Thomson heat. What other technique could be used?

8.5   Which of the semiconductors in Table 8.1 are *p*-type and which *n*-type? Explain your answer.

8.6   Consider the data of Table 8.1, particularly the positive thermoelectric powers for Cu, Ag, and W. What does this imply about our simple model of the band structure of a conductor? What modifications must be made?

8.7   Referring to Table 8.1, what signs would you expect the Hall coefficients of the semiconductors and conductors to have?

8.8   Calculate (using Table 8.1 for data) the ideal Peltier heat for a junction of *p*-type ZnSb and *n*-type ZnSb at 200°C. Referring to Figures 8.5 to 8.9, we see that no direct *p-n* contacts are made. Metal is used as an intermediary for both hot and cold junctions. Why are direct *p-n* contacts avoided? Why are *p*- and *n*-type versions of the same material often used in the same thermocouple?

8.9   Calculate (using Table 8.1 for data) the ideal thermoelectric voltage developed by the ZnSb thermocouple described in Problem 8.8, with the high-temperature junction at 200°C and the low-temperature junction at 0°C. Suppose a (Li$_{0.05}$Ni$_{0.95}$O) thermocouple with hot junction at 1100°C were used. What would its thermoelectric voltage be?

8.10   Considering the Peltier heats and thermoelectric powers only, calculate the ideal reversible efficiency of a thermoelectric generator, with hot junction at T_2 and cold junction at T_1. The work output is the thermoelectric voltage times the charge passed, and the net heat input is the *net* Peltier heat of the thermocouple, per unit charge passed. Compare the efficiency (W/Q_2) with that of the ideal reversible heat engine described in Section 1.4 of Volume II.

8.11   The heat-flow loss in a thermoelectric generator is found to be $0.7 \times 10^{-3}(T_2\text{-}T_1)$ watts. The internal resistance is found to be $2 \times 10^6\,\Omega$, so that the electrical power loss is that number times the square of the current. The generator consists of 300 thermocouples, with the cold side at 0°C. The generator may be built of: (a) Al-Ag thermocouples with hot sides at 100°C, (b) ZnSb *p-n* thermocouples with hot sides at 200°C, and (c) (Li$_{0.05}$Ni$_{0.95}$O) *p-n* thermocouples with hot sides at 1100°.
Data on these materials may be found in Table 8.1. The hot sides have 50 watts delivered to them in total, by an oil stove. The generator must deliver 0.1 amp.

Make a crude calculation to compare the performances of the three types of thermocouples, (a), (b), and (c):

(1)   Subtract the heat loss in each case from the 50 watt input.

(2)   Multiply the result by the ideal efficiency to get the available electric work output. Assume the ideal efficiency to be $W/Q_2 = (T_2\text{-}T_1)/T_2$. (See Problem 8.10.)

(3)   Subtract the electrical power loss at 0.1 amp from the above. (a) How much power is left for use in each case? (b) What is the voltage onput of each at zero current? At 0.1 amp? At maximum power output? (c) Calculate and compare the figures of merit. Use the data of Table 8.1; assume S, σ_T, and σ_E to be independent of temperature.

8.12   For the three thermocouples mentioned in the previous problem, calculate optimum efficiencies by using Equation 8.12 and Table 8.1. Why do these efficiencies not agree with the answers to part 3(a) of the previous problem?

8.13   Outline a brief plan for a research program to develop improved thermoelectric materials. Indicate the techniques to be used and the areas of crystal structure, etc., to be investigated.

8.14 List possible applications for thermoelectricity in the under-developed countries.

8.15 Compare the efficiency of various processes of converting heat to electrical energy: photocells, thermoelectrics, dynamos, and (optional) magnetohydrodynamics.

8.16 Derive the Kelvin relations (Equations 8.05 and 8.06) in the following manner:

(a) Pass one coulomb of charge through the thermocouple of Figure 8.1. The work done is then V_{12}, on the thermocouple. Considering this work, together with the Peltier heats at the two junctions and the Thomson heats, apply the First and Second Laws of Thermodynamics, obtaining two equations. When calculating the entropy of the Thomson heat, use the average temperature of each thermocouple, $\frac{1}{2}(T_2 + T_1)$.

(b) Let $T_2 = T_1 + \Delta T$, and let ΔT go to zero, to yield, for instance,

$$\lim_{\Delta T \to 0} \left[\frac{\pi_{12}(T + \Delta T) - \pi_{12}}{\Delta T} \right] = \frac{d\pi_{12}}{dT}.$$

Solve for the Kelvin relations. Is this derivation correct? Why? Or why not?

CHAPTER NINE

Magnetism

The magnetic properties of solids originate in the motion of the electrons and in the permanent magnetic moments of the atoms and electrons. Diamagnetism, which is very weak, arises from changes in the atomic orbital states induced by the applied field. Paramagnetism is the result of the presence of permanent atomic or electronic magnetic moments. Ferromagnetism, which is very strong, occurs when quantum mechanical exchange interactions align adjacent magnetic moments in the same direction. If the exchange interaction aligns the moments in opposite directions and only one type of moment is present, cancellation occurs, and the material is called *antiferromagnetic*. If two or more types of moment are present, there is a net moment equal to the difference and the material is called *ferrimagnetic*. Above some critical temperature, a ferro-, antiferro-, or ferrimagnetic material becomes paramagnetic. Ferro- and ferrimagnetic materials consist of *domains* or regions of completely magnetized material, separated by boundaries. The domain structure is determined by the various types of magnetic energies, such as the isotropic exchange energy, the magnetostatic energy, the magnetocrystalline and the magnetostrictive energies.

9.1 INTRODUCTION

In a uniform magnetic field $\mathbf{B}$, the torque τ on a magnetic dipole of dipole moment $\mathbf{p}_m$ is

$$\tau = \mathbf{p}_m \times \mathbf{B} \tag{9.1}$$

the vector product (or cross product) of $\mathbf{p}_m$ and $\mathbf{B}$. Bar magnets and any circulating currents have dipole moments. A small loop of current I which encloses the area A has the dipole moment

$$\mathbf{p}_m = I\mathbf{A} \qquad (9.2)$$

where the vector $\mathbf{A}$ has a magnitude equal to the enclosed area and a direction normal to the area. If the loop has N turns, the right side of Equation 9.2 must be multiplied by N. A bar magnet also has a dipole moment. The field surrounding a bar magnet is also similar as Figure 9.1 indicates. By analogy with the electric dipole, which consists of two opposite and separated charges, it is

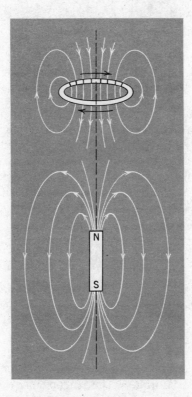

Figure 9.1   The magnetic field distributions around a current loop and a bar magnet. The north and south poles are indicated on the latter.

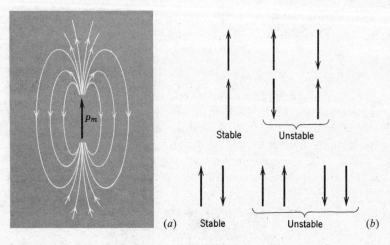

Figure 9.2   (a) The field of a magnetic dipole such as the current loop or bar magnet of Figure 9.1.   (b) Stable and unstable positions for a pair of dipoles.

sometimes useful to imagine the existence of *magnetic poles*.   In the case of a bar magnet with two poles of magnitude $+m$ and $-m$ separated by distance l the magnetic moment is ml.   We cannot separate the poles of the bar magnet, for instance, by breaking the magnet in half: two smaller dipoles are formed instead.   Diligent searching by many has failed to uncover any particle having the properties of a magnetic *monopole*.

To obtain the potential energy of the dipole in the magnetic field it is necessary to integrate Equation 9.1

$$U = -\mathbf{p}_m \cdot \mathbf{B} + \text{const.} \tag{9.3}$$

From Equation 9.3 it is possible to obtain the force on the dipole. For one dimension the force is

$$F_x = \mathbf{p}_m \cdot \frac{d\mathbf{B}}{dx} \tag{9.4}$$

Thus, in homogeneous fields, there is no net force on the dipole yet there can be a torque.   The reaction of two dipoles on one another can be calculated from Equation 9.1 or 9.3.   Figure 9.2*b* illustrates stable and unstable arrangements.

9.2   MAGNETIZATION

In free space, the induction **B** is related to the field strength **H** by

$$\mathbf{B} = \mu_0\mathbf{H}, \tag{9.5}$$

where $\mu_0 = 4\pi \times 10^{-7}$ henry/meter; μ_0 is called the permeability of vacuum.   In a solid material,

$$\mathbf{B} = \mu\mathbf{H}, \tag{9.6}$$

where μ is in general not equal to μ_0.   Equation 9.6 can alternatively be expressed as

$$\mathbf{B} \equiv \mu_0(\mathbf{H} + \mathbf{M}) = \mu\mathbf{H} \tag{9.7}$$

M is called the *magnetization* of the solid.   Thus the solid is somehow responsible for the appearance of the extra magnetic induction field $\mu_0\mathbf{M}$, in addition to the free-space induction $\mu_0\mathbf{H}$.   It is possible to show (see Problem 9.2) that **M** is just the density of magnetic dipole moment, that is, the dipole moment per unit volume.   The force on a small piece of material of volume V and magnetization **M** can, therefore, be obtained by considering the dipole moment of the entire piece of material, $V\mathbf{M}$, and applying Equation 9.4;

$$F_x = V\left(\mathbf{M} \cdot \frac{\partial \mathbf{B}}{\partial x}\right). \tag{9.8}$$

The magnetization of a solid may be regarded as resulting from the appearance of dipole moments in the solid when immersed in a magnetic field.

The magnetization of a solid can also be described in alternate terms.   The ratio μ/μ_0 is called the *relative permeability* μ_r.

$$\mu_r \equiv \frac{\mu}{\mu_0} \tag{9.9a}$$

Since the magnetization is proportional to the applied field, the factor of proportionality called the *susceptibility* is given by

$$\chi_m = \frac{H}{M} \tag{9.9b}$$

hence

$$\mu_r = 1 + \chi_m = \frac{\mu}{\mu_0} \qquad (9.10)$$

We need only know one parameter, either M, χ_m, μ_r, or μ, as a function of H, to specify the rest.

9.3   MAGNETIC UNITS

The units used above are in the rationalized mks system. H and M are measured in amperes per meter, and B in webers per square meter. The permeability μ has the dimensions webers per ampere-meter, or *henries* per meter. The susceptibility and relative permeability are dimensionless. The potential and current are in volts and amperes. In research and publications it is more usual to employ gaussian units. To convert from mks to

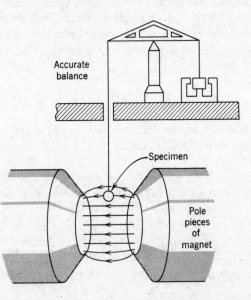

Accurate
balance

Specimen

Pole
pieces
of
magnet

Figure 9.3   Schematic of a magnetic balance for measuring magnetization. The force is proportional to $M(\partial B/\partial z)$. Here, $\partial B/\partial z$ is in a vertical direction.

gaussian units, the following rules are useful:

H.   Multiply amperes/meter by $4\pi \times 10^{-3}$ to obtain *oersteds.*
B.   Multiply webers/(meter)^2 by 10^4 to obtain *gauss.*
M.   Multiply amperes/meter by 10^{-3} to obtain *oersteds.*

In the gaussian system, the permeability of a vacuum $\mu_0 = 1$, so that Equations 9.6 and 9.7 become

$$\mathbf{B} = \mu\mathbf{H} = \mathbf{H} + 4\pi\mathbf{M} \qquad (9.11)$$

The magnetization M of a material can be calculated from a direct measurement by using Equation 9.8.   Figure 9.3 illustrates a simple method.   The specimen is placed in a known field with a small gradient, $\partial B_z/\partial z$, along the axis of the specimen.   If the specimen is cylindrical, with its axis parallel to the applied field, its magnetization will not distort the field lines.   If B_z and $\partial B_z/\partial z$ are known, and the force on the specimen F_z is carefully measured by the balance, Equation 9.8 gives M.   Other techniques for measuring magnetization are described in the Bibliography at the end of the chapter.

9.4   DIAMAGNETISM

When the magnetization is negative, we say that the solid is called *diamagnetic.*   In diamagnetic solids such as Bi, Cu, Ag, and Au, the magnetic moments oppose the $\mathbf{B}$, causing the field inside the solid to be less than the field outside.   According to Lenz's law, a current induced by a changing field will always oppose the change that induces it.   Thus, the field of the induced current opposes the change in the applied field.   Any conductor will temporarily oppose penetration of a magnetic field by the flow of its conduction electrons, but these currents die out very quickly.   More permanent changes occur in the orbital motions of electrons bound to the atoms.   These changes also follow Lenz's law, and lead to a true diamagnetism.   Diamagnetism in most solids is very weak, with susceptibilities of the order of 10^{-5}.   It can be observed only when all the other types of magnetism are totally absent.   Diamagnetism is not important in engineering applications.

9.5   PARAMAGNETISM

Many solids that have small but positive magnetic susceptibilities are called *paramagnetic*. The source of paramagnetism is the permanent magnetic moments possessed by the atoms and electrons in the solid. Equation 9.3 indicates that a magnetic dipole can minimize its potential energy by lining up with the magnetic field. Such alignment makes a positive contribution to the magnetization. The atoms in a solid may have permanent magnetic moments. As noted in Chapter 1 of Volume I, there is a magnetic moment, $l\mu_B$, associated with an electron's atomic quantum state, where l is the azimuthal quantum number of the state, and μ_B is called the *Bohr magneton;* it is equal to $eh/4\pi m_e$ (9.27×10^{-24} amperes per square meter). The component of the magnetic moment parallel to the applied field is also quantized and is equal to $m_l\mu_B$ where m_l is the *magnetic quantum number,* which may have any integral value between $\pm l$.

It is convenient to regard the orbital magnetic moment of the atomic electron as resulting from its motion about the nucleus. This is analogous to the magnetic moment of the circulating current described by Equation 9.2. In any solid, then, the orbital motions of the atomic electrons gives rise to permanent atomic magnetic moments. In the case of filled atomic shells, however, all the moments cancel.

A second source of permanent moments is the *electron spin.* We should not envision the electron spin in the same way as we do the orbital motion because the electron spin is a relativistic effect and is not analogous to the motion of a spinning body. The term *spin* is used to refer to the fact that electrons possess intrinsic angular moments and magnetic moments like those expected of charged body. The spin angular momentum of the electron is $\pm\frac{1}{2}(h/2\pi)$, and the spin magnetic moment is $\pm\mu_B$. The spin angular momentum and magnetic moment can be aligned *up* or *down*. Since any quantum state can be filled by two electrons of opposite spin, the net spin magnetic moment of a completely filled state is zero. The net spin magnetic moment of a completely full atomic shell is also zero. Open shells may fill in a complicated manner, but if we know which states are occupied and how, we may be able to combine the spin and orbital moments and obtain the net atomic magnetic moment. A third contribution, of the order of 10^{-3} Bohr magnetons per atom arises from the spin of the atomic nucleus. It can only be observed by sensitive measurement.

Since the atoms in a solid may have permanent magnetic moments, we can consider the behavior of these moments by assuming that they are free to rotate. The potential energy of a single dipole in a magnetic field is given by Equation 9.3 as $-p_m B \cos \theta$. It is a function of the angle between the dipole and the field. The dipole tends to lower its energy by lining up with the applied field and therefore gives rise to a positive susceptibility, χ_m. However, it is necessary to reckon with the effect of temperature. According to the Maxwell-Boltzmann distribution, the number of dipoles having energy, E, is proportional to $e^{-E/kT}$. Employing Equation 9.3 this expression becomes $e^{-\frac{p_m \cdot B}{kT}}$. At temperature T, this assumes that the dipoles make different angles with the field, with the probable number at any angle proportional to the Boltzmann factor. More dipoles are aligned with, rather than against, the field. Hence a small positive susceptibility results. If we rigorously apply Equation 9.3 and the Maxwell-Boltzmann distribution, we can calculate the paramagnetic susceptibility of N dipoles of moment p_m (see Problem 9.8), and obtain

$$\chi_m = \frac{M}{H} \cong \frac{\mu_0 N p_m{}^2}{3kT} = \frac{C}{T} \tag{9.12}$$

where μ_0 is the permeability of free space. Equation 9.12, called the Langevin equation, is correct for p_m of about a Bohr magneton, H less than 10^6 amp/meter, and T at or above room temperature. The linear dependence of the susceptibility χ_m on the reciprocal of the temperature is called the Curie law, and the constant C in Equation 9.12, the Curie constant.

The Langevin equation does not apply to conduction electrons for they obey the Fermi-Dirac distribution. The use of Equation 9.3 and the Fermi-Dirac distribution gives a paramagnetic susceptibility for the conduction electrons which is 50 per cent larger than the observed susceptibility. The reason is that the applied field also changes the states of the conduction electrons in a manner similar to the diamagnetic changes in the atomic states.

A separate diamagnetic contribution equal to $\frac{1}{3}$ the paramagnetic contribution, arises which cancels the excess predicted. The resulting net susceptibility for N conduction electrons is

$$\chi_m = \frac{\mu_0 N \mu_B{}^2}{E_F} \qquad (9.13)$$

where E_F is the Fermi level. This type of paramagnetism is independent of temperature. Generally, the susceptibility of paramagnetic solids is between 10^{-3} and 10^{-5}. Paramagnetism usually masks the atomic diamagnetism present in solids.

9.6 FERROMAGNETISM

The transition metals Fe, Co, and Ni, rare earth metals such as Gd, and a few oxides such as CrO_2 and ErO display very large magnetizations. In fact, these substances remain magnetized even when the field is removed. Their magnetization is not reversible; that is, it depends on how the field is applied. The magnetization curve for a ferromagnetic material is shown schematically in Figure 9.4. As the applied field, H is increased B begins to increase slowly. Then, the slope rises sharply and B rapidly increases until the *saturation induction, B_s*, is attained. With further increase in the field, the slope levels off. Upon decreasing the field, the original curve is not retraced. At H equal to zero, the specimen is still magnetized and $B = B_r$, the *remanent induction*. At $B = 0$, $H = H_c$ *the coercive force*. If H is now made negative and the specimen saturated in the reverse direction before returning to zero field, the symmetric curve shown in Figure 9.4 is obtained with a saturation, coercive force, and *remanence* equal to those on the positive side. Such irreversible, double-valued *hysteresis* behavior is characteristic of the magnetic behavior of ferromagnetic materials. The work required to go around the *hysteresis loop* once is proportional to the area enclosed by the curve. If H is brought back to zero and the cycle repeated at less than saturation, a similar hysteresis curve of smaller area is obtained.

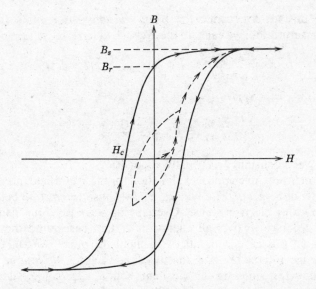

Figure 9.4 *B* versus *H* hysteresis loop for a ferromagnetic material. The dashed line indicates the behavior on initially increasing the applied field *H*. If the field is increased to saturation, and then decreased to zero and increased in the reverse direction, the solid curve will result. The dotted curve represents a hysteresis curve which does not reach saturation.

9.7 WEISS FIELD AND MAGNETIC DOMAINS

The first question one might ask about ferromagnetic materials is the source of the large magnetization. The saturation magnetization is so large that virtually all of the magnetic dipoles in a ferromagnetic material must be lined up with the field. Our previous discussion of paramagnetism indicated that such behavior would not occur, due to the disruption of the alignments by thermal energy, if $p_m \cdot B$ were the total energy of the dipole. To make our model fit the facts, it is necessary to add a term to the energy which will make the dipoles all line up. By adding a new energy term we are actually making the same assumption as Weiss (1907) since he was the first to postulate the existence of such a field. It is given by

$$H_W \equiv \lambda M \qquad (9.14)$$

The constant λ is called the *Weiss constant*. For applied fields below saturation, we can add the Weiss field to H, in Equation 9.12,

$$M \equiv \frac{\mu_0 N p_m^2}{3kT}(H + \lambda M) = \frac{C}{T}(H + \lambda M) \qquad (9.15)$$

and solve for (M/H):

$$\frac{M}{H} = \chi_m = \frac{C}{T - \lambda C} \qquad (9.16)$$

The Weiss constant λ is usually about 10^3.

A quantum mechanical explanation for the Weiss field, proposed by Heisenberg (1928), involves an *exchange interaction* between neighboring electron spins. Overlapping wave functions can lead to a decrease in over-all energy in certain cases, and therefore favor, in this case, a parallel alignment of spins. An analogous exchange interaction was discussed in Chapter 1, Volume I, to account for bonding in covalent solids. In some materials (antiferromagnetic) the exchange energy leads to an opposite or *antiparallel* spin alignment.

We must now consider the hysteresis curve (Figure 9.4) for a ferromagnetic material. The question arises as to why a ferromagnetic material may be magnetized *or* unmagnetized at zero field. It is necessary to explain why B and M depend on *how* H is applied [for any value of H up to saturation, B (and, therefore, M) may have one of many values, depending on how we have gone around the curve].

To provide an explanation for the hysteresis effects observed in ferromagnetic materials. Weiss (1907) offered a second novel idea: magnetic domains. A ferromagnetic material is divided up into small regions, each of which is at all times completely magnetized. The situation is represented by Figure 9.5 for an unmagnetized material. Although each domain is magnetized, the material as a whole will have zero magnetization, as in the figure. To give the material a net magnetization, one direction must predominate in the domains. There are two possible ways to magnetize a domain structure. The most obvious is to permit rotation of the individual domain magnetization. However, less energy is required if domains initially parallel to the applied field grow at the expense of their less favorably oriented neighbors.

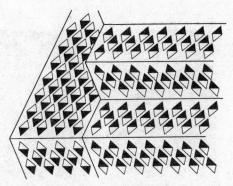

Figure 9.5   Domains in a demagnetized ferromagnetic solid. All of the moments in each domain are aligned, but the size, shape, and direction of magnetization of each domain is such that the net magnetization is zero.

In Figure 9.6 the latter concept is used to explain the magnetization curve.   Initially, no domain growth occurs as the field H is increased.   Then the favorably oriented domains grow and the magnetic induction B increases rapidly.   Finally, domain

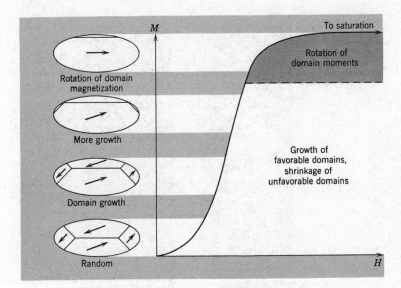

Figure 9.6   Domain growth and rotation in a ferromagnetic material and the associated magnetization curve M versus H.

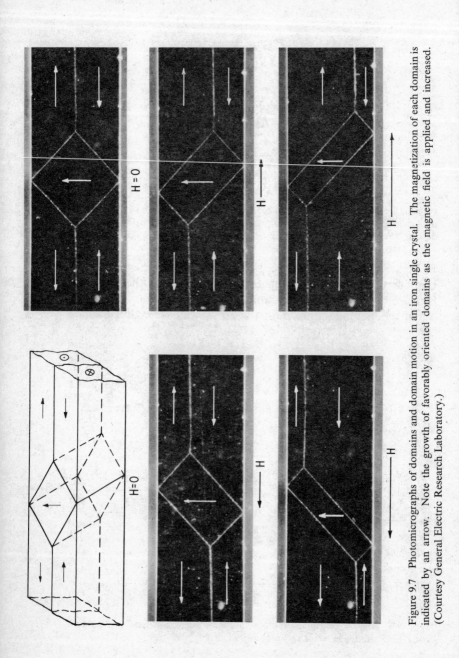

Figure 9.7 Photomicrographs of domains and domain motion in an iron single crystal. The magnetization of each domain is indicated by an arrow. Note the growth of favorably oriented domains as the magnetic field is applied and increased. (Courtesy General Electric Research Laboratory.)

growth stops as we enter the saturation region and rotation of the remaining unfavorably aligned domains occurs. Since domain rotation requires higher energy than domain growth the slope of the B versus H curve decreases. When the field is removed, the specimen remains magnetized. Although the domains tend to rotate back, the large aligned domains do not easily revert to the original random arrangement. If a reverse field $(-H)$ is applied, the domain structure may be changed to produce a resultant zero magnetic induction. The magnitude of the applied field required is equal to coercive force H_c. Once magnetized, the state $H = 0$, $B = 0$ is no longer attainable by simply changing the applied field.

The domain structure postulated by Weiss in 1907 was actually observed more than 25 years later using colloidal iron oxide to delineate domain boundaries. The photomicrographs shown in Figure 9.7 were obtained using the Bitter (1931) colloidal iron oxide technique. They illustrate the growth of favorable domains in iron single crystals when the field is increased.

In ferromagnetic materials thermal energy can overcome the Weiss field at some high temperature called the Curie temperature. Above their Curie temperatures, ferromagnetic materials become paramagnetic.

9.8   ANTIFERROMAGNETISM AND FERRIMAGNETISM

There are two other important classes of magnetic behavior. As mentioned in the last section the exchange energy sometimes has the opposite effect to ferromagnetism: the spins are aligned oppositely rather than in the same direction. This phenomenon is known as *antiferromagnetism*. The susceptibility is then positive and increases as the temperature increases since the thermal energy disrupts the antiparallel moment arrangements and presents them to line up with the field. Figure 9.8 compares the alignment of magnetic moments and temperature effects on ferro- and anti-ferromagnetic materials. The peak in the χ_m versus T curve shown in Figure 9.8b for antiferromagnetic material, is called the *Neel temperature* Θ_N and corresponds to the Curie temperature in ferromagnetic materials.

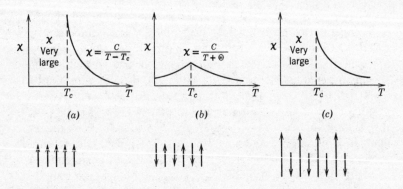

Figure 9.8 Magnetic susceptibility versus T for (a) ferromagnetic, (b) antiferro-magnetic, and (c) ferrimagnetic materials, with magnetic moment alignments indicated for each case.

A certain class of complex oxides, called *spinels,* having the composition $XOFe_2O_3$ (where X is a metal), exhibit ferromagnetic interactions yet have antiparallel spins arranged as in Figure 9.8c. Although nearest neighbors have opposing spins in such *ferrimagnetic* material, one set of spin moments is larger than the other as shown. Consequently a net moment results. The magnetization of the ferrimagnetic spinels, known as *ferrites,* is large enough to be useful commercially although the saturation value is not as large as in the ferromagnetic case. Ferrimagnetic materials have hysteresis curves like Figure 9.4, and domain structures similar to those in ferromagnetic materials.

9.9 FERROMAGNETIC ANISOTROPY AND MAGNETOSTRICTION

The B versus H curve of a single crystal of ferromagnetic material depends on the crystal orientation relative to the field. Figure 9.9 shows B versus H curves for single crystal of iron and nickel, with the applied field along $<111>$, $<110>$, and $<100>$ directions. In each case, one direction is more easily magnetized than the others. Saturation occurs at a lower field in this direction. For BCC iron, the direction of *easy magnetization* is

$<100>$, and *hard magnetization* is $<111>$. In FCC nickel, the reverse is true. On this account the different grains in a poly-crystalline material approach saturation differently. The grains whose directions of easy magnetization lie near the field direction saturate at low applied fields; those whose directions of hard magnetization lie near the applied field direction rotate their moment into the field direction only at higher fields. The work required to rotate all the domains against the anisotropy is called *magnetocrystalline energy*.

When the magnetic electron spin dipole moments of the atoms in a solid are rotated into alignment, the length of the bonds between the atoms changes. The fields of the dipoles themselves affect the atomic spacing, as they may attract or repel each other. Therefore, the shape and volume of a ferromagnetic solid changes as it is magnetized. The principal change, called the *magnetostriction*, is a reversible strain along the axis of magnetization. Depending upon the material, the solid may expand or contract.

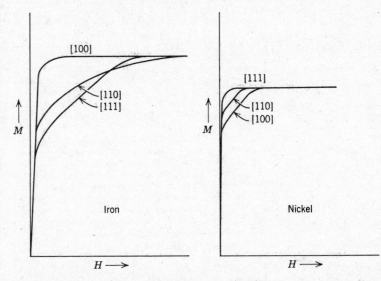

Figure 9.9 Magnetization versus applied field for field direction along the [100], [110], and [111] directions of iron and nickel single crystals. [After K. Honda and S. Kaya, *Sci. Rep. Tôhoku Imp. Univ.*, Ser. 1, Vol. 15, p. 721 (1926) for the iron, and S. Kaya, *Sci. Rep. Tôhoku Imp. Univ.*, Ser. 1, Vol. 17, pp. 639, 1157 (1928) for the nickel.]

The magnetostriction is anisotropic, not only because the magnetization curve is anisotropic, but also because the elastic properties of the crystal are anisotropic.   For any given crystal direction, the magnostriction approaches a constant value at high magnetic fields.   Magnetostriction and magnetization usually saturate at the same time.   Figure 9.10 shows magnostriction versus applied field curves for nickel and iron single crystals at $<100>$, $<110>$, and $<111>$ field orientations.   The strain $\delta l/l$ along the field axis is plotted as a function of the field intensity in Figure 9.10b.

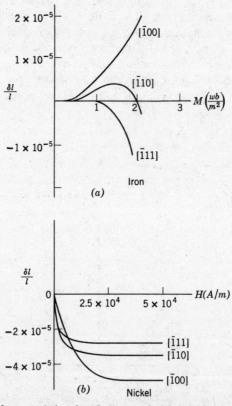

Figure 9.10   Magnetostriction data for iron and nickel single crystals showing the anisotropy.   [After W. L. Webster, *Proc. Roy. Soc. (London)*, Ser. A, Vol. 109, p. 570 (1925) for iron, and Y. Masiyama, *Sci. Rep. Tôhoku Imp. Univ.*, Ser. 1, Vol. 17, p. 945 (1928) for nickel.]

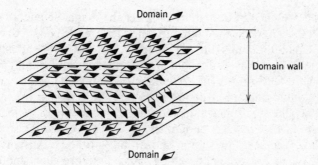

Domain wall

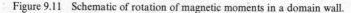

Figure 9.11   Schematic of rotation of magnetic moments in a domain wall.

9.10   MAGNETIC ENERGY AND DOMAIN STRUCTURE

There are four types of energy which should be considered in any discussion of domain structure. The energy of the interaction which makes adjacent dipoles line up is called the exchange energy. The energy resulting from the dipole moment of the material is called the *magnetostatic energy*. The anisotropy of magnetization must also be considered, because some crystall orientations require higher fields to achieve the same magnetization. Because magnetostriction involves mechanical work the *magnetostrictive* energy must be included.

The domain boundary where the magnetic moments change from one orientation to another is analogous to a grain boundary, where crystal orientation changes. In contrast with a grain boundary the orientation changes gradually across a domain boundary, as shown in Figure 9.11. The reason for the gradual transition is that the exchange interaction is strong enough to force the moments into relatively close alignment. An abrupt change in orientation would require large amounts of energy. The domain boundary has a *surface energy* which arises from the exchange interaction energy of the misalignments. A further contribution to the surface energy results because the boundary has moments aligned in directions of hard magnetization. The presence of a domain boundary, therefore, raises the energy of a ferromagnetic or ferrimagnetic material by effects which are similar to those of interphase or grain boundaries in chemical systems.

The presence of domains can lower the magnetic energy of the system, even when the boundary energies are included. Consider the single-domain solid of Figure 9.12*a*, with its surrounding dipole field. Any reduction in the intensity and extent of the magnetic field results in a reduction of the magnetostatic energy. The double-domain structure of Figure 9.12*b* lowers the magnetostatic energy somewhat, and the introduction of *domains of closure* in Figure 9.12*c* eliminates it. Since no poles exist on the surface, no field lines originate at the surface. Although the decrease in magnetostatic energy may be more than enough to offset the increase in energy due to the additional domain bound-

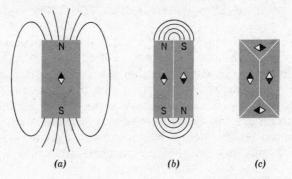

(*a*)   (*b*)   (*c*)

(*a*) One domain. (*b*) Reduction of magnetostatic energy by splitting into two domains. (*c*) Further reduction in magnetostatic energy by addition of domains of closure.

(*d*)   (*e*)

(*d*) Exaggeration of the tendency of the magnetostriction (negative, here) to pull apart the domain boundaries. (*e*) Lessening of magnetostrictive stresses by appearance of smaller domains.

Figure 9.12   The evolution of domain shape by successive reduction of magnetostatic and magnetostrictive energies, which more than offset the domain wall energy added.

aries, it is also necessary to consider anisotropy and magnetostriction. If Figure 9.12 were concerned with, say, nickel, all the domains could be magnetized in the easy $<100>$ directions. Some anisotropy energy would be unavoidably stored in the domain boundaries. The magnetostriction would, if negative, tend to pull the specimen apart, as shown in Figure 9.12d. Due to different geometry and direction of magnetization magnetostriction causes a different *net* boundary displacement in each domain. To make the displacements at the boundaries smaller, the domains must become smaller, as shown in Figure 9.12e. The domain size thus depends on the balance between the energies of magnetostriction and the domain boundaries.

DEFINITIONS

Magnetic Dipole Moment of a Current Loop ($\mathbf{p}_m$). For a one turn current loop $\mathbf{p}_m = I\mathbf{A}$. The direction of $\mathbf{A}$ is related to that of the current by the right-hand rule: if I flows in the direction of the fingers, the thumb points in the direction of $\mathbf{A}$. $\mathbf{p}_m$ is expressed in amp-meter^2.

Magnetization (M). Defined by the relation $\mathbf{B} \equiv \mu_0(\mathbf{H} + \mathbf{M})$; M is equal to the dipole moment density, i.e., total moment per unit volume.

Relative Permeability μ_r: Defined by the relation $\mu_r \equiv \mu/\mu_0$, where μ is the permeability of the material and μ_0 is the permeability of free space ($4\pi \times 10^{-7}$ henry/meter).

Magnetic Susceptibility (χ_m). Defined by the relation $M = \chi_m H$.

Electron Spin. A term used to refer to the fact that electrons have permanent angular moments and magnetic moments.

Diamagnetism. The magnetic property of materials for which M (and therefore χ_m) is negative. The relative permeability μ_r is less than one.

Bohr Magneton (μ_B). The fundamental quantum of magnetic moment: 9.27×10^{-24} amp-meter^2.

Exchange Energy. The energy associated with the quantum mechanical coupling that aligns the individual atomic dipoles within a single domain.

Crystal Anisotropy Energy. The energy of magnetization which is a function of crystal orientation. The difference in energy between the hard [111] direction and the easy [100] in Fe is about 1.4×10^4 joule/m^3.

Domain Wall Energy. The sum of contributions from exchange and crystalline anisotropy energy in the domain wall region.

Magnetostatic Energy. When a ferromagnetic substance produces an external field, magnetic potential energy or magnetostatic energy is present.

Paramagnetism. The small positive susceptibility due to the weak inter-action and independent alignment of permanent atomic and electronic magnetic moments with the applied field.

Magnetic Permeability (μ). Defined by the relation $\mathbf{B} = \mu\mathbf{H}$.

Ferromagnetism. The appearance of a very large magnetization due to the parallel alignment of neighboring magnetic moments by an exchange interaction.

Antiferromagnetism. The opposite alignment of adjacent atomic magnetic moments in a solid produced by the exchange interaction.

Ferrimagnetism. A special case of antiferromagnetism, where the opposed moments are of different magnitudes and a large net magnetization thereby results.

Domain. A region in a ferro- or ferrimagnetic material where all the moments are aligned.

Magnetostriction. The change in length along the direction of magnetiza-tion of a multidomain solid.

Magnetostrictive Energy. The energy due to the mechanical stresses generated by magnetostriction in the domains.

Langevin Equation. For paramagnetic materials, $\chi_m = (\mu_0 N p_m^2)/3kT$

Curie Law. For paramagnetic materials, $\chi_m = C/T$

Curie Constant (C). The proportionality constant between χ_m and $1/T$ in the Curie law.

Curie Temperature. If a ferromagnetic material is heated above its Curie temperature, it becomes paramagnetic.

Weiss Field (H_W). A hypothetical internal magnetic field, strong enough to make the spin magnetic moments in a solid line up despite the effect of thermal energy.

Weiss Constant (λ). The proportionality constant between the Weiss Field and the magnetization: $H_W = \lambda M$.

Saturation (M_s or B_s). The maximum value of magnetization M_s or mag-netic induction B_s for a ferromagnetic material.

Coercive Force (H_c). The applied field required to reduce the induction of a magnetized material to zero.

Remanence (B_r). The value of B in the specimen when H is reduced to zero.

Hysteresis. The irreversible B-H characteristic of ferromagnetic and ferrimagnetic materials.

Neel Temperature Θ_N. The temperature above which an antiferromagnetic or ferrimagnetic material becomes paramagnetic.

BIBLIOGRAPHY

SUPPLEMENTARY READING:

Halliday, D., and Resnick, R., *Physics for Students of Science and Engineering,* Wiley, New York, 1965, Chapters 33 and 37.

Hutchison, T. S., and Baird, D. C., *The Physics of Engineering Solids,* Wiley, New York, 1963, Chapter 12.

Nesbitt, E. A., *Ferromagnetic Domains,* Bell Telephone Laboratories, Murray Hill, New Jersey, 1962.

Stanley, J. K., *Electrical and Magnetic Properties of Metals,* American Society for Metals, Metals Park, Ohio, 1963, Chapter 5.

ADVANCED READING

Bozorth, R. M., *Ferromagnetism,* Van Nostrand, New York, 1951.

Chikazumi, S., and Charap, S. H., *Physics of Magnetism,* Wiley, New York, 1964.

Bates, L. F., *Modern Magnetism,* Cambridge University Press, Cambridge, 1951.

PROBLEMS

9.1   Diamagnetic specimens are repelled by the pole of a bar magnet, while paramagnetic specimens are weakly attracted, and ferromagnetic specimens strongly attracted.   Explain by comparing the magnetizations and using Equation 9.4.

9.2   A needle of a diamagnetic material is freely suspended at its center of mass and allowed to rotate in a uniform magnetic field.   In what position relative to the field will the needle come to rest if the needle is (a) paramagnetic and (b) ferromagnetic?

9.3   The dipole moment of an iron atom is 1.8×10^{-23} amp-meter^2. The density of iron is 7.87 g/cc, and the atomic weight is 55.85.   Avogadro's number is 6.023×10^{23}.

(a) What is the dipole moment of a completely magnetized, i.e., saturated iron bar 10 cm long and 1×1 cm square, parallel to the long axis of the bar?

(b) Suppose that the dipole moment of part *a* were permanently fixed in the bar, parallel to the long axis.   What torque would be required to hold the bar perpendicular to an applied magnetic field of 50,000 gauss.

9.4   It has been suggested that the BCC structure is stabilized by an exchange interaction whereby the total electron spin of each atom is opposite to those of each and every one of its nearest neighbors.   Show that such an arrangement is possible in BCC, but not in HCP or FCC.

9.5   In traveling around the hysteresis curve, is work done?   How would you measure the work done from the hysteresis curve, and how would you determine it experimentally?

9.6   List and describe briefly the different techniques for measuring the strength of a magnetic field (use the library).

9.7   List and describe briefly the different techniques for measuring the magnetization and magnetic susceptibility (use the library).

9.8   Derive Langevin's equation for paramagnetism. A dipole (p_m) at the angle θ to the field contributes $p_m \cos \theta$ to the magnetization. The number of dipoles between θ and $\theta + d\theta$ is proportional to $(\sin \theta e^{\mu_0 p_m H \cos \theta / kT}) d\theta$. Integrate the total contribution, and derive the Langevin equation. To eliminate the constant of proportionality, remember that the total number of dipoles is a constant, say N.

9.9   Compare the relative magnitude and the sign of μ, χ_m, and μ_T for (a) diamagnetic, (b) paramagnetic, and (c) ferromagnetic material.

9.10   The susceptibility of a paramagnetic material is positive and decreases as T increases. Why should the susceptibility of an antiferromagnetic material also be positive, yet increase as T increases?

9.11   Consider an electron orbiting about a nucleus with angular velocity ω, in a circular path of radius r. Show that the ratio of orbital magnetic moment to orbital angular momentum is $-e/2m$. (The minus sign comes from the negative charge of the electron.) This ratio is called the gyromagnetic ratio.

9.12   If the angular momentum of the electron is quantized in units of $h/2\pi$, show from the result of Problem 9.11 that the magnetic moment is quantized in Bohr magnetons. Does the reverse hold true?

9.13   Referring to Problem 9.11 show that the gyromagnetic ratio of any rotating distribution of electrical charge must be $\pm e/2m$. Why should the gyromagnetic ratio for the electron spin be equal to e/m?

9.14   Suppose material of susceptibility χ_m is subjected to a field of tensity H. What are the values of B and H inside the material? Suppose a small cylindrical cavity of cross-section area A and length l were in the middle of the material, with the axis of the cylinder parallel to H. What are the values of B and H inside the cavity?

9.15   How much current would have to flow around the cylindrical cavity of Problem 9.14 in order to make H in the cavity equal H in the material? What is the dipole moment of that current? (This is also dipole moment of the material that was removed to make the cavity.) Show from this that M is equal to the dipole moment per unit volume present in the material.

9.16   (Library problem) (a) Indicate graphically the dependence of exchange energy on interatomic separation for Mn, Cr, Fe, Co and Ni.

(b) Now explain why a Cr-Mn-Al (Heusler alloy) is ferromagnetic.

9.17   Consider the electronic structure of Fe, Co and Ni atoms. How many 4s and 3d electrons has each of these atoms? Using a box configuration, show how three spins could be coupled to give a magnetic atomic moment of $4\mu_B$ for Fe, $3\mu_B$ for Co and $2\mu_B$ for Ni. Account for the fact that the measured values found are 2.22, 1.7 and 0.61, respectively, instead of the above.

9.18   (Library Problem) Describe ferromagnetic resonance. What parameters are measured, and what type of information is sought?

9.19   (Library Problem) Describe nuclear magnetic resonance. What is measured, and what type of information is sought?

9.20   At any boundary, including a domain boundary, the component of B perpendicular to the boundary must have the same value on both sides, and, if no true currents flow, the component of H tangential to the boundary must likewise be continuous. Assuming all the domains in Figure 9.8 to be completely magnetized, show that Figure 9.8 obeys these boundary conditions. What sort of geometry is favored by the boundary conditions?

9.21   Describe in a short essay the Bitter technique for delineating domain boundaries in ferromagnetic materials. Use Nesbitt, E. A., as a reference.

CHAPTER TEN

Magnetic Materials

Soft magnetic materials are processed to have maximum permeability and low hysteresis and eddy current losses. Structural defects may increase the coercive force and hysteresis by pinning domain walls. Silicon-iron, the most common soft magnetic material, may be significantly improved by developing a preferred orientation in the easy magnetization direction. However, it has relatively low initial permeability and high eddy current losses at audio frequencies due to low resistivity. In these respects, the nickel-iron alloys are superior and their properties can be further improved by magnetic annealing. At high frequencies, nickel-iron alloys become inadequate due to energy losses, and are supplanted by ferrites and, at very high (e.g., microwave) frequencies, by garnets, both of which have high resistivities and consequently low eddy current losses.

Alnicos are the most popular hard magnetic material due to their high energy product. Heat treatment of Alnico alloys in a magnetic field increases the energy product considerably. In both soft and hard alloys, magnetic annealing induces an anisotropy with improved properties in the desired direction. Permanent magnets with high energy products may be made of single domain pure iron or iron-cobalt powders compacted in a magnetic field.

10.1 INTRODUCTION

A very important part of the behavior of ferromagnetic and ferrimagnetic materials depends on the microstructure, the phase constitution, and the structural imperfections present. In diamagnetic and paramagnetic materials, the susceptibility is of primary interest and structure plays a secondary role, influencing the susceptibility only slightly. In ferromagnetic materials, the saturation magnetization, B_s, is primarily a function of composition, as it depends on some average atomic moment. The permeability, coercive force, and the general shape of the hysteresis curve are, however, very sensitive to the structure, and thus to the thermal, chemical, mechanical, and magnetic history of the material.

The shape of the hysteresis curve of a magnetic material is of paramount engineering importance. The largest use of magnetic materials is in the cores of transformers, motors, inductors, and generators. A ferromagnetic core in an inductor or transformer increases the flux linkage many times over, due to its magnetization, and the device may then handle much more energy. However, some of the energy is invariably lost due to the magnetic hysteresis discussed in the previous chapter. In a transformer which handles 60-cycle current, the core traverses the entire hysteresis loop 60 times per second; and, of course, the problem of energy loss becomes much more serious at higher frequencies. The work required to cycle repeatedly over the hysteresis loop is wasted, since it is only used to push domain walls back and forth as the magnetization of the core changes. To minimize the energy wastage, the area under the hysteresis loop should be as small as possible, but the permeability and saturation induction should be as large as possible, as illustrated in Figure 10.1. Materials having small hysteresis losses are called *soft* magnetic materials.

Another type of a-c energy loss stems from Faraday's law. The fluctuating magnetic field induces the flow of electrical currents (called *eddy currents*) in the magnetic core, and the energy is dissipated by Joule heating of the core, that is, electrical resistance heating. Since the voltages induced are determined by the rate of change of the field in the core, which is fixed by the frequency of the current handled by the transformer or other device, we can reduce the eddy current losses only by increasing the electrical resistivity of the core which will, in turn, reduce the flow of induced current. It is not generally possible to increase substantially the resistivity without increasing the hysteresis loss. Some relief is possible by using cores of laminated or powdered material to restrict current flow. Even so, silicon-iron alloys

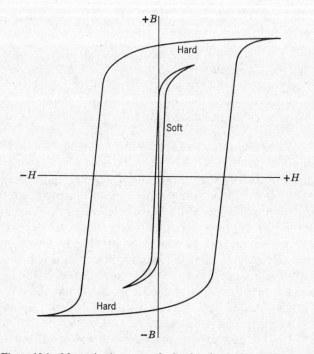

Figure 10.1   Magnetization curves for hard and soft magnetic materials.

which are used at power frequencies (e.g., 60 cycles) show too high a loss at audio frequencies where nickel-iron alloys (Permalloys) must be used.   At higher frequencies, the ceramic ferrites which have higher electrical resistivities supplant Permalloy.

While materials for a-c applications must be easy to magnetize and demagnetize, there is also a practical need for materials of high coercivity.   In permanent magnets, a large area under the hysteresis curve a high coercive force and a large saturation magnetization, as shown in Figure 10.1 are desired.

10.2   HARD AND SOFT MAGNETIC MATERIALS

The process of magnetization of a ferro- or ferrimagnetic material consists, as mentioned previously, of first moving the domain walls so that favorably oriented domains grow and unfavorable domains shrink.   If the domain walls are easy to move, the coercive force is low, and the material is easy to magnetize, it is called a soft magnet.   If it is difficult to move the domain walls, the coercive force is large, and the material is magnetically *hard*. In a very hard material, the domain walls may be totally immobile. The immobilization of the domain walls may be caused by structural defects, particularly nonmagnetic inclusions, voids or precipitates of a nonmagnetic phase.   Domain walls are attracted to these imperfections because the wall energy and the magnetostatic energy is thereby reduced.   As Figure 10.2 shows, a spherical nonmagnetic inclusion in the middle of a domain has one half of its area abutting on south poles of the magnetic moments, and the other half on north poles.   Since the poles attract, a more favorably energetic situation becomes possible by reduction of the distance between the north and south regions. Two ways to reduce the magnetostatic energy are shown in Figure 10.2*b*.   The inclusion may lie in a domain boundary, thereby

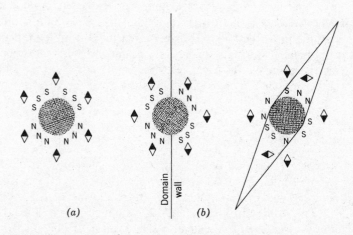

(*a*)   (*b*)

(*a*) The distribution of north and south poles about the nonmagnetic inclusion has a high magnetic energy due to the separation of the poles

(*b*) The north and south poles may be brought closer together by mixing them up; this occurs if the particle lies in a domain boundary or its own blade-shaped domain, as shown above

Figure 10.2   The attraction of domain walls to nonmagnetic inclusions; blade-shaped domains.

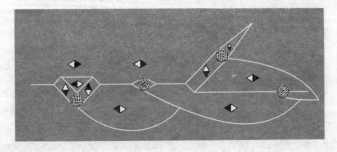

Figure 10.3 The pinning of domain walls by nonmagnetic inclusions and their associated domains.

eliminating part of the boundary as well as bringing the moments closer together. The inclusion may also have a domain of its own, characteristically blade-shaped, as shown in Figure 10.2*b*. The blade-shaped region distributes the poles over a longer area. Combinations of the two cases usually occur as shown in Figure 10.3. Thus, the presence of impurities, precipitates, plastic strain, or any other crystalline imperfections are expected to lead to magnetic hysteresis due to the immobilization of domain boundaries by the imperfections. Defects which make a material magnetically hard also usually make it mechanically hard. Although defects increase the electrical resistivity and thus decrease eddy current losses, they increase hysteresis losses.

10.3 IRON-SILICON ALLOYS

The most widely used type of soft magnetic material is the iron-silicon alloy. Before 1900, ordinary low-carbon steel was used for low-frequency power applications, in transformers, generators, and motors. Today iron-silicon alloys, which cut power losses by a factor of three, are used instead. The addition of silicon to iron increases the electrical resistivity, thus reducing eddy current losses and hysteresis. It also increases the magnetic permeability. The presence of silicon in iron makes rolling into sheet (most transformer cores are made of laminated sheet) more difficult. In the melting of silicon-iron in the electric arc furnace, undesirable sulfur, phosphorous, nitrogen, carbon and oxygen are removed. Rolled material can be further refined by annealing in moist, followed by dry, hydrogen to eliminate residual carbon and oxygen. Such processing improves the permeability.

If the rolling and annealing of silicon-iron sheet is carefully controlled, preferred crystal orientation can be induced. The directions of easy magnetization then lie in the rolling direction as illustrated in Figure 10.4. It is easier to magnetize the *textured* sheets shown in Figures 10.4b and 10.4c in the rolling direction than randomly textured sheet (Figure 10.4a), for the unfavorably oriented grains of Figure 10.4a require higher magnetizing fields.

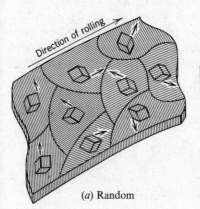

(a) Random

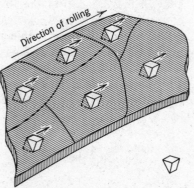

(b) (110)[001] texture: (110) plane parallel to plane of sheet, [001] direction parallel to rolling direction

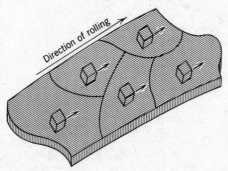

(c) (100)[001] texture: (100) plane parallel to plane of sheet, [001] direction parallel to rolling direction

Figure 10.4   Random and preferred orientations in polycrystalline silicon-iron sheet. The small cubes indicate the orientation of each grain.

Figure 10.5 illustrates the great advantage of random-textured silicon iron over plain cast iron (or plain carbon steel), and the even greater advantage of textured silicon-iron over the random form, in permeability. Lower hysteresis also accompanies easier saturation.

10.4   SOFT IRON-NICKEL ALLOYS

The initial magnetization behavior of the iron and iron-silicon alloys is magnified in Figure 10.5. It shows that the permeability of these materials in weak fields is relatively low. Low initial permeability is no real problem in power equipment where core materials are operated at high magnetizations. For high sensitivity and fidelity in communications equipment, iron-silicon alloys are, however, not suitable. Iron-nickel alloys are usually used in such equipment. Table 10.1 compares the important

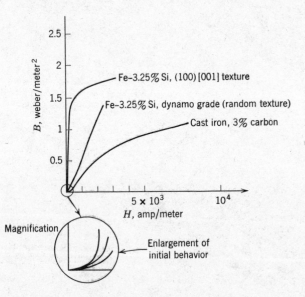

Figure 10.5   A comparison of the initial B versus H curves of previously demagnetized cast iron, random-textured silicon-iron, and cube-textured silicon-iron. The enlargement of the origin shows the relatively low initial permeability of these alloys.

Table 10.1   Typical Magnetic Properties of Various Soft Magnetic Materials

MATERIAL	INITIAL RELATIVE PERMEABILITY (μ_r AT $B \sim 0$)	HYSTERESIS LOSS JOULE/m^3 PER CYCLE	SATURATION INDUCTION, WEBER/m^2
Commercial iron ingot	250	500	2.16
Fe-4% Si, random	500	50–150	1.95
Fe-3% Si, oriented	15,000	35–140	2.0
45 Permalloy (45% Ni-55% Fe)	2,700	120	1.6
Mumetal (75% Ni-5% Cu- 2% Cr-18% Fe)	30,000	20	0.8
Supermalloy (79% Ni- 15% Fe-5% Mo-0.5% Ma)	100,000	2	0.79

properties of iron and iron-silicon with three commercial iron-nickel alloys. The Permalloys and Mumetals have higher initial permeability, lower hysteresis and eddy current losses. These advantages are gained, as Table 10.1 indicates, at the expense of saturation induction, but are, nonetheless, useful for operation at audio and low radio frequencies.

The permeabilities of the iron-nickel alloys are sensitive to heat and mechanical treatment, especially for compositions between 50 and 80 per cent nickel. If such an alloy is slowly cooled from above 600°C to below 400°C, its permeability is about one half that of the same alloy rapidly cooled through this temperature range. This behavior arises from the order-disorder transformation in the nickel-iron system. Below 500°C, the equilibrium structure of such FCC alloys contains Ni atoms at face centers and Fe atoms at face corners. The perfectly ordered structure has a relatively low permeability; slow cooling past 500°C favors ordering. Quenching on the other hand suppresses the order transformation, and leads to higher permeability. Higher permeabilities can be attained by magnetic annealing; this requires cooling from 600°C in a magnetic field. The alloy develops a strong magnetic anisotropy with the direction of easy magnetization the same as the direction of the applied field during the cooling, as shown in Figure 10.6.   Magnetic annealing also affects the hysteresis curve, as shown in Figure 10.7 (next page).   The *square* hysteresis loop that results is desired in computer, magnetic amplifier, and pulse transformer applications.   Extreme care is necessary in handling the iron-nickel alloys, since plastic strain drastically lowers the permeability.   Permalloy sheet or tape is therefore laminated or wound into the desired form and subsequently annealed.   Such processing can raise the relative permeability of *supermalloy* (Table 10.1) to over one million.

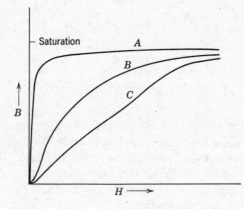

Figure 10.6   Magnetization anisotropy for a 21.5% Fe − 78.5% Ni alloy.   The alloy was cooled from 600°C in a magnetic field.   Curve *A* is the *B-H* curve with the field applied parallel to the direction of the field during cooling; curve *C* is the *B-H* curve with the field perpendicular.   Curve *B* is for the same alloy, cooled in the absence of a magnetic field.   (From *Physics of Magnetism,* S. Chikazumi and S. H. Charap, Wiley, New York, 1964, page 360.)

10.5   SOFT FERRITES AND GARNETS

The hysteresis curves, domain structure, and domain motion of ferrites is similar to those of ferromagnetic metals.   As mentioned in Chapter 9, their large magnetic permeability is due to an antiferromagnetic interaction which rigidly aligns neighboring magnetic moments in opposite directions.   Since one set of moments is larger than the other, a net moment results.   Most ferrites have

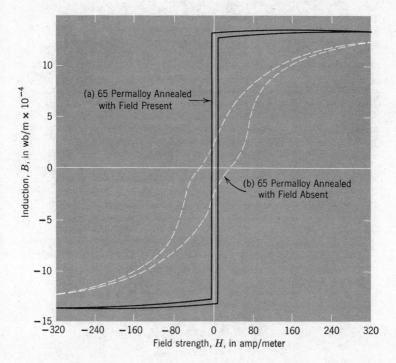

Figure 10.7   The effect of magnetic annealing on the hysteresis curve of a 65% Ni — 35% Fe alloy.   (After R. M. Bozorth, *Ferromagnetism*, Van Nostrand, New York, 1951, p. 121.)

the *inverse spinel* structure shown in Figure 10.8.   The oxygen atoms form an FCC lattice; the iron atoms are equally divided between octahedral sites, and one set of tetrahedral sites.   The divalent atoms fill the remaining set of tetrahedral sites.   In the ferrimagnetic ferrites the moments of all tetrahedrally coordinated ions are opposed to those of the octahedrally coordinated ions by antiferromagnetic interaction.   Thus the set of iron moments in the octahedral sites cancel those in the tetrahedral sites.   The residual moment is that of the divalent ions in the remaining tetrahedral sites.   It is possible to predict the saturation magnetization if the moments of the divalent ions are known.   There are hard and soft ferrites, just as in the case of ferromagnetic materials.   Soft ferrites have lower saturation induction than soft ferromag-

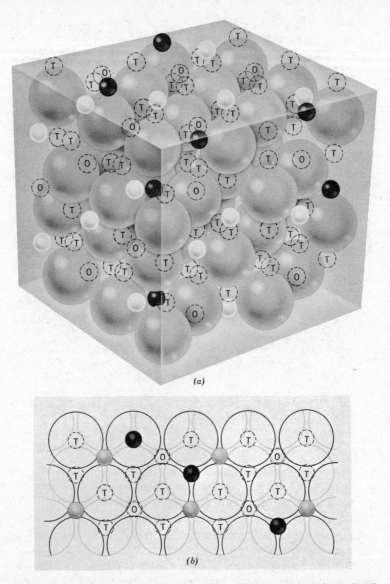

Figure 10.8 (a) The spinel structure; large gray spheres are oxygen ions. Small light spheres are filled octahedral sites. Small dark spheres are filled tetrahedral sites. Normal spinels have trivalent ions in octahedral sites, divalent ions in tetrahedral sites. Inverted spinels, which include the ferrites are ferrimagnetic and have trivalent iron ions divided evenly between tetrahedral and octahedral sites; the divalent ions lie in tetrahedral sites. Open circles indicate empty sites; O for octahedral and T for tetrahedral. (b) The packing sequence for the spinel structure: two close-packed oxygen planes, with the interstitial ions and sites between.

netics, yet higher resistivities. Consequently, eddy current losses are much lower; they are usually less than one millionth of those in typical silicon-irons. At frequencies above 10^6 cycles per second, ferrites cores are mandatory.

Ferrites are carefully made by mixing powdered oxides, compacting, and sintering at elevated temperatures. The high-frequency transformer in television and fm receivers are almost always made with ferrite cores. Many ferrites, such as (50% MgO, 50% MnO)-Fe_2O_3, have square hysteresis loops and are, therefore, useful in computers. The nickel-zinc ferrites, such as the Ferroxcube listed in Table 10.2, can be magnetically annealed to improve the squareness of the hysteresis loop. Ferrites with large magnetostrictive effects, are sometimes used in electromechanical transducers. In high frequency applications, magnetostriction in ferrites can lead to undesirable noise and even failure.

The net magnetic moment of the ferrite unit cell, as stated above, is equal to the magnetic moment of the divalent metal ions. The saturation magnetization and, usually the permeability can be increased by divalent ions of larger magnetic moment. The rare earth ions with large moments are mainly trivalent and too large in volume to fit the tetrahedral interstices of the ferrite structure. The ferrimagnetic oxides called garnets can, however, incorporate the rare-earth ions in their structure. The formula for ferrimagnetic garnets is $3M_2O_3$-$5Fe_2O_3$, where M is any rare-earth element (Sm, Eu, Gd, etc.). The garnet, $3Y_2O_3$-$5Fe_2O_3$, called yttrium iron garnet, or simply YIG, has a high resistivity and very low hysteresis loss at microwave at frequencies.

Table 10.2 *Typical Magnetic Properties of Soft Ferrites*

MATERIAL	INITIAL RELATIVE PERMEABILITY	SATURATION INDUCTION, WEBER/m^2	ELECTRICAL RESISTIVITY Ω-m
Ferroxcube A[4] (48% MnO-Fe_2O_3, 52% ZnO-Fe_2O_3)	1200	0.36	0.5×10^6
Ferroxcube B' (36% NiO-Fe_2O_3, 64% ZnO-Fe_2O_3)	650	0.29	10^3
NiO-Fe_2O_3	17	0.23	10^3

10.6   HARD MAGNETS

Materials having the highest possible saturation magnetization, remanence, and coercive force are used as permanent magnets. Figure 10.9 compares the demagnetization curves of some of the better commercial permanent magnet materials. Alnico 5 contains 14% Ni, 24% Co, 8% Al, 3% Cu, and the rest Fe. Other members of the Alnico (Al-Ni-Co) family also contain Ti. Ferroxdur, $BaFe_{12}O_{19}$, is a magnetically hard ferrite. The oxygen atoms form an HCP lattice with Ba dissolved substitutionally. The Fe atoms fit in the interstitial sites.

The usefulness of a permanent magnet is determined by the magnetic energy it can deliver at various flux densities. Since the magnetic potential energy of the magnetized material is approximately equal to $BH/2$, the available energy may be obtained by plotting (BH) as a function of H along the demagnetization curve as shown in Figure 10.10; this curve is called an external energy curve. The maximum value of $BH/2$ is called the *energy product* of the magnet. Figure 10.9 gives the energy products of various materials with their demagnetization curves. In general, the

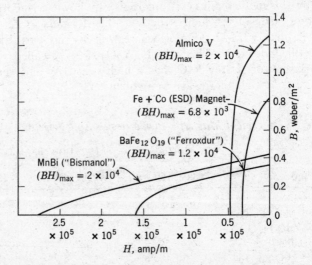

Figure 10.9   Demagnetization curves for hard magnetic materials. (After Bozorth from *The Science of Engineering Materials,* V. Golldman ed., Wiley, 1957.)

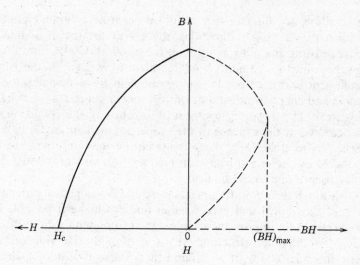

Figure 10.10   The external energy curve (BH) of Alnico is on the right.   It has been plotted from the demagnetization curve on the left.   The maximum value of ($\frac{1}{2}BH$) is called the energy product.

energy available from any magnet is approximately equal to the volume of the magnet times its energy product.

Hard magnetic alloys are classed into three groups, dependent upon the microstructural transformation which gives the high energy product.   Transformation hardening alloys undergo a martensitic transformation upon cooling with a resultant fine-scale structure of high mechanical hardness and internal stress. High carbon steels, and alloy steels containing W, Cr, Co or Al, fall into this category.   Precipitation hardening alloys demonstrate a yet finer structure, with a high resistance to domain growth and rotation.   In the group are Cunife (Cu-Ni-Fe), Cunico (Cu-Ni-Co), Alnico alloys, and Silmanal (Ag-Mn-Al).   Although Silmanal has a relatively low saturation magnetization, its coercive force is considerably greater than Bismanal, shown in Figure 10.9.   The order hardening alloys typified by FePt and CoPt form superlattices upon cooling, and again exhibit a high coercive force.

The Alnico type alloys are commercially the most important of the hard magnetic materials.   Large magnets are made by special casting techniques and small ones by powder metallurgy.   If the cast alloys are directionally solidified to grow columnar grains with parallel $<100>$ directions, the energy product is doubled. After casting the alloy is solution annealed at 1300°C and then given a short (~10 min.) heat treatment at 800°C. If this heat treatment is carried out in a magnetic field the demagnetization curve and energy product are considerably enhanced, as shown in Figure 10.11. This improvement in direction of the applied field is acquired at the expense of the properties at right angles to the field. After the 800°C heat treatment further improvement is possible by prolonged heat treatment (~14 hours) at ~580°C in the absence of a magnetic field.

Although the above treatments were first developed empirically, recent theoretical and experimental findings make it possible to associate these improvements in magnetization with subtle alterations in the microstructure. By use of high resolving power electron microscopy, it was found that Alnico type alloys, which are single phase BCC at 1300°C, decompose into two separate BCC cubic phases α and α' at 800°C. The α' precipitate is rich in Fe and Co and therefore has a higher magnetization than the

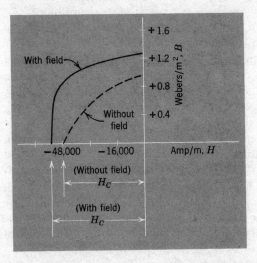

Figure 10.11 The effect of magnetic annealing on the demagnetization curve of Alnico V. The heat treatments were at 800°C with and without the magnetic field, followed in both cases by a 600°C heat treatment. The magnetic field directions during annealing and demagnetization are the same.

Figure 10.12   An electron micrograph, magnification about 100,000×, of an Al-Ni-Co-Fe alloy after a 800°C heat treatment without magnetic field.   The light-colored network is the Ni-Al rich α phase, which has precipitated on the {100} planes. The dark material within the net is α', the Fe-Co rich phase.   (Courtesy K. J. de Vos.)

Ni-Al rich α phase.   The α phase precipitates rapidly on {100} planes at 800°C as indicated in Figure 10.12.   If the 800°C heat treatment is carried out in a magnetic field, the α phase appears to prefer the field direction.   This leaves the higher magnetization α' precipitate as fine elongated particles within an α network which is evident in the electron micrograph of Figure 10.13.   The magnetic characteristics of the material after this heat treatment are attributed to the shape anisotropy of the α' phase.   The difference in microstructure between material heat-treated with and without a magnetic field can be seen by comparing Figures 10.12 and 10.13.

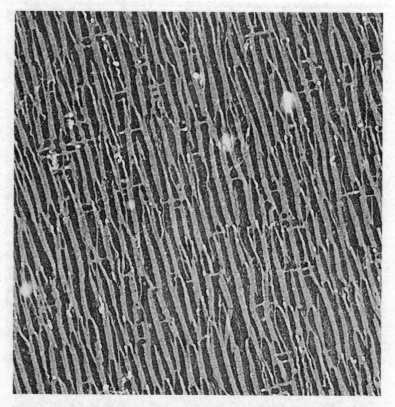

Figure 10.13 An electron micrograph, magnification about 100,000×, of a Al-Ni-Co-Fe alloy after a 800°C heat treatment in an applied magnetic field. The α phase is light, the α' phase dark. The α' phase has elongated in the direction of the applied field resulting in anisotropy of the coercive force. (Courtesy K. J. de Vos.)

It is difficult to find any change in microstructure after the 14-hour heat treatment at 580°C. The further increase in coercivity is probably due to transfer by diffusion of Fe and Co from the α to the α' phase. This tends to make the α phase paramagnetic.

The Alnico-type magnets can be viewed as an agglomerate of elongated high magnetization α' precipitates, smaller in thickness than a domain wall boundary. The precipitates are separated from one another by a paramagnetic α phase network. Heat-treated Alnico has a high energy product because it is difficult to alter the magnetic moments of the α' precipitates.

10.7 FINE PARTICLE MAGNETS

Hard magnetic material of high coercive force can alternatively be made by bonding together magnetic particles smaller than one domain wall thickness. Since the particles are single domains and separated from one another by a resin or nonmagnetic metal binder, the magnetization can only be changed by moment rotation. Ferroxdur, the hard magnetic ferrite discussed above, is prepared by magnetic pressing, sintering and magnetic annealing of the fine BaO-ferrite powder.

Fine particle magnets with energy products approaching that of Alnico have been produced from single domain Fe or Fe-Co alloy particles bonded with Pb. The fine particles, precipitated electrolytically from a solution on a liquid mercury cathode, are elongated along the easy direction of magnetization, and less than a domain boundary width (about 150 Å) in thickness. The high coercive force of these elongated single domain (ESD) magnets, results from three factors: the fine, single domain particle, the shape anisotropy of the particles, and the fact that the particle is elongated in the easy magnetization direction. The material is cooled, the binder hardens, and a material with an energy product of 9×10^3 joules/m^3 results under good conditions. Figure 10.9 shows the demagnetization curve for an iron-cobalt ESD material. The cobalt is added to improve saturation magnetization, and energy products of 1.5×10^4 joules/m^3 have been achieved with Fe-Co ESD materials. For iron, the theoretical ESD energy product under optimum conditions is 1.6×10^5 joules/m^3, but this assumes zero binder volume and completely solid iron, that is, a 100 per cent dense powder. If necessary (the alloying elements for Alnico are strategic materials), better ESD materials could be made, but at present, ESD material is relatively expensive and seldom used in place of Alnico.

10.8 TAPES AND FILMS

Oriented silicon-iron and iron-nickel alloy sheet as described previously is used for transformer applications at low frequencies. Thinner sheet called tape is required for frequencies greater than 20 cps. At one megacycle the laminations used are below 10^{-5}

meters. Such tapes are used in magnetic amplifiers for computers and automatic controls. To produce even thinner magnetic components Ni-Fe films, ≤ 1000 Å thick, are evaporated, sputtered or plated in a magnetic field. These are used in magnetic memory storage devices. In common recording tape elongated fine particles of γFe_2O_3 are oriented on a plastic tape to produce a rectangular hysteresis loop of high coercive force. For fidelity in sound reproduction, it is necessary that the residual magnetization of such tape be proportional to the initial electrical signal produced by sound. Therefore, the core material should be as soft and as loss-free as possible. To this end, the soft ferrites are preferable to Permalloys, and ferrite cores with switching times of the order of 10^{-6} seconds are a mainstay of the computer builder. Magnetic thin films, ~ 3000 Å thick, are even faster: 10^{-8}–10^{-9} seconds. The films are usually of Permalloys and can be prepared by electroplating, decomposition of chemical vapors, or direct evaporation and deposition in vacuum. Domain motion does not occur in these films; the moments actually rotate directly and, therefore, great rapidity of magnetization is possible. (Actually, semiconducting devices switch even faster. Esaki diodes can switch in 10^{-11} to 10^{-12} seconds.)

DEFINITIONS

Hysteresis Loss. The work dissipated in tracing a *B-H* loop; essentially the work needed to move the domain boundaries during a single magnetization cycle.

Eddy Current Loss. Power loss of magnetic materials in alternating fields due to induced currents in the material.

Soft Magnetic Material. A higher permeability material having low hysteresis and low coercive force.

Hard Magnetic Material. A material having high coercive force, and high saturation magnetization.

Spinel ($MgAl_2O_4$). A compound whose Al (trivalent) atoms occupy octahedral sites, and Mg (divalent) atoms occupy tetrahedral sites in an FCC oxygen lattice. Other trivalent and divalent metal ions may be used in place of the Al and Mg, respectively. $ZnFe_2O_4$ is an example.

Inverted Spinel. The same oxygen lattice as a spinel, with half the trivalent ions on tetrahedral sites. The ferrimagnetic ferrites are inverted spinels.

Silicon iron.   Fe-3 or 4% Si; a soft magnetic material with higher saturation magnetization used in motors, low frequency generators, and power transformers.

Permalloys.   Ni-Fe alloys with high initial permeabilities used where high sensitivity is needed.

Directions of Easy Magnetization.   The crystallographic direction in a ferro-magnetic single crystal which is most easily magnetized.

Preferred Orientation (Texture).   Crystallographic alignment of the gram structure in a polycrystalline solid.

Alnicos.   A most widely used class of permanent magnet alloys, which contain Fe, Al, Ni, Co, and Cu or Ti.

Energy Product.   The highest magnitude of the product of B and H to be found on the demagnetization curve of a hard magnetic material.

ESD (Elongated Single-Domain) Magnet.   A magnet made of aligned fine particles which are below the domain wall size and elongated in their direction of easy magnetization.

Ferrite.   Material whose formula is $MO \cdot Fe_2O_3$, where M is a divalent metal, whose structure is of the spinel family.   The ferrimagnetic ferrites have the inverted spinel structure.

Garnet.   A group of oxides similar to the spinel.   The ferrimagnetic garnets have the formula $2M_2O_3\text{-}5Fe_2O_3$, where M is a trivalent rare-earth metal.

Magnetic anneal.   The heat treatment of a magnetic material in a magnetic field which induces an anisotropy by aligning various features of the microstructure with the field, e.g., directional ordering in Permalloys, the shape of the α' phase in Alnico 5, and the alignment of the powder particles in an ESD magnet.

BIBLIOGRAPHY

SUPPLEMENTARY READING:

Bozorth, R. M., "Ferromagnetism," pp. 253–291 of *Recent Advances in Science,* New York University Press-Interscience, New York, 1956.   Reprinted in Bell Telephone System Monograph No. 2679.

Bozorth, R. M., "The Physics of Magnetic Materials," pp. 302–335 of *The Science of Engineering Materials,* Ed. by J. E. Goldman, Wiley, New York, 1957.

Brailsford, F., *Magnetic Materials,* Methuen-Wiley, New York, 3rd Ed., Revised 1960.

Stanley, J. K., *Electrical and Magnetic Properties of Metals,* American Society for Metals, Metals Park, Ohio, 1963, Chapters 5–7.

ADVANCED READING:

Bates, L. F., *Modern Magnetism,* Cambridge Univ. Press, New York, 1961 (4th edition).

Bozorth, R. M., *Ferromagnetism,* Van Nostrand, New York, 1951.

Chikazumi, S., and Charap, S. H., *Physics of Magnetism,* Wiley, New York, 1964.

Hoselitz, K., *Ferromagnetic Properties of Metals and Alloys,* Oxford University Press, London, 1952.

PROBLEMS

10.1   What sort of magnetic properties are desirable in:

(a) A 60 cycle per second power transformer?

(b) An audio-frequency transformer?

(c) A pulse transformer?

(d) A radio-frequency transformer?

(e) A *magnetic shield?*

10.2   (Library Problem) Describe the operation of a *magnetic amplifier,* also called a *saturable reactor.* What sort of magnetic material is required?

10.3   Permalloy cores for computer logic and memory elements are wound from Permalloy tape and then annealed afterwards, rather than before. Why? What property is important?

10.4   The extra elements (copper, molybdenum, etc., see Table 10.2) are added to the basic nickel-iron Permalloy composition to raise the electrical resistivity and, more important, to slow the order-disorder transformation. This is essential to the manufacture of large items. Why?

10.5   Mumetal (see Table 10.2) is used as magnetic shielding, that is, to shield various items from low-level magnetic fields. Why? How does Mumetal or any other high permeability alloy act as a magnetic shield? As an example, draw the flux lines around a hollow cylinder of high permeability material.

10.6   Estimate the energy product in the following:

(a) Alnico 5 has a B_r of 1.2 webers/m² and an H_c of 5×10^5 amp/m².

(b) For soft iron, B_r is 1.5 webers/m² and H_c is 75 amp/m.

(c) For $BaFe_{12}O_{19}$ (barium ferrite), H_c is 1.5×10^5 amp/m and B_r is 0.38 webers/m².

10.7   Materials with high coercive force, generally, have relatively low initial permeabilities. Explain why you would expect this, in terms of domain wall motion.

10.8   Power transformers emit a loud hum when they operate. The cores are usually silicon iron sheet laminations. What are possible sources of the noise?

10.9   Many potential applications exist for magnetic core materials in devices which would be irradiated by neutrons and other high-energy particles. The main problem is that the radiation raises hysteresis losses sharply. Why? How would you deal with this problem?

10.10   Consider the rare-earth metals whose atomic numbers run from 58 to 71 (Ce to Lu). Refer to the periodic table for the atomic structures, and estimate the atomic magnetic moments, in Bohr magnetons. Compare with the ionic and metallic moments which may be found on page 448 of the Chikazumi reference, cited in the Bibliography of this Chapter.

10.11   Are all the octahedral and tetrahedral sites in the ferrite structure filled? If not, what would determine the manner in which the sites were filled? (A review of the atomic packing in Chapter 2, Volume I, may help).

10.12   Compare the magnetization of pure bulk iron with the pure iron of an ESD magnet. Does the actual microscopic process of magnetization differ? In what way, if any?

10.13   Laminated silicon iron sheet is commonly used for transformer cores, in preference for castings or other shapes. There are at least two reasons for this practice. Name them.

10.14   There are eight divalent nickel ions present per unit cell of nickel ferrite, $NiO \cdot Fe_2O_3$. From the atomic structure of nickel, what would you expect the magnetic moment of the nickel ion to be? The edge of the unit cell of nickel ferrite is 8.34 A.U. long. (The measured value is 0.34 webers/meter^2.) If cobalt ferrite and iron ferrite (also called magnetite) have substantially the same unit cell sizes, what would their saturation magnetization be?

10.15   Zinc ions, which are divalent, have no magnetic moment. What is the saturation magnetization of zinc ferrite? Actually, zinc ferrite has the normal spinel structure: zinc ions in tetrahedral sites and iron in octahedral sites. If zinc ferrite is added to any magnetic ferrite such as nickel or cobalt, the saturation magnetization of the magnetic ferrite is raised. Why? (Remember, magnetic ferrites have the inverted spinel structure: divalent ions in tetrahedral sites, with half the iron ions; the remainder of the iron in the octahedral sites.)

10.16   Austenite, the face-centered cubic form of iron, is paramagnetic, as is austenitic stainless steel if it is quenched from 1000°C. However, if austenitic stainless is cold rolled or otherwise severely deformed, it becomes ferromagnetic. If it is slowly cooled from 1000°C, austenitic stainless shows traces of ferromagnetism. Explain.

10.17   In view of your answer to Problem 10.16, what precautions would you take if you were responsible for the manufacture of totally nonferromagnetic instruments from 18-8 austenitic stainless? Assume that your material is purchased through the usual industrial outlets.

CHAPTER ELEVEN

Superconducting Materials

A superconducting material has zero electrical resistivity if the *critical temperature, critical magnetic field,* and *critical current density* are not exceeded. The critical temperature is a function of the position of the superconductor in the periodic table. Superconductors display two distinct types of magnetic behavior. The *ideal Type I superconductor* completely excludes the magnetic field until the critical field is exceeded. The *ideal Type II superconductor* behaves similarly at low fields, but then allows a gradual field penetration, and returns to the normal state when penetration is complete. The magnetic field does not have a discontinuity at the surface of a Type I superconductor, but decays to zero in a short distance, called the *penetration depth*. The critical field can be related thermodynamically to the free energy difference between the normal and superconducting phases. Thin films, of thickness less than the penetration depth, have high critical fields due to their reduced magnetization. The difference in magnetic properties between Type I and Type II superconductors can be explained by the existence of a positive (Type I) and *negative* (Type II) *surface energy* between the superconducting and normal phases. Nonideal Type II superconductors can carry high supercurrent densities in high magnetic fields. The high critical current density is related to the presence of structural imperfections. Very high magnetic fields have been produced by superconducting solenoids.

11.1 INTRODUCTION

If a specimen of mercury is cooled below $4.15°K$, all of its electrical resistivity vanishes. This phenomenon is called superconductivity, and the temperature of the transition is referred to as the critical temperature T_c for solid mercury. In Figure 11.1 the

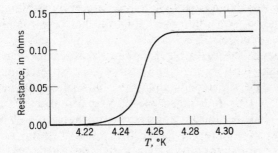

Figure 11.1 Electrical resistivity of mercury as a function of temperature. (H. Kamerlingh Onnes, Leiden Comm. Vol. 122b, 1911.)

resistance is plotted as a function of temperature for a specimen of mercury, near T_c, as determined in 1911 by H. Kammerlingh Onnes. There are at least 24 other superconducting elements, and a much larger number of superconducting compounds and alloys, with critical temperatures ranging from 1° to nearly 19°K. Table 11.1 lists the critical temperatures of all the known super-

Table 11.1 Periodic Table Indicating Superconducting Elements and Transition Temperatures (The Elements in Shaded Boxes are Type II Superconductors; further Purification May Show that They Are Type I)

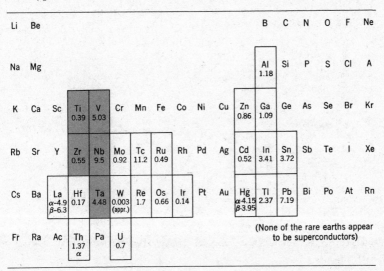

conducting metals on a periodic table. Recently strontium titanate and doped germanium were found to have critical temperatures of the order of $0.01°K$.

It is important to realize that the most sensitive measurements one can make indicate that the resistivity of a superconductor is truly zero in the superconducting state. Measurements made on lead at $4.2°K$ (the T_c for lead is 7.19) with a sensitivity of 10^{-25} ohm-meter fail to indicate any resistivity whatever. By way of comparison, very pure copper, with a resistivity ratio of about 20,000, has a resistivity of about 7×10^{-3} ohm-meters at $4.2°K$.

11.2   CRITICAL FIELD AND CURRENT DENSITY

Superconductivity will disappear if the temperature of the specimen is raised above its T_c or if a sufficiently strong magnetic field or current density is employed. The applied field necessary

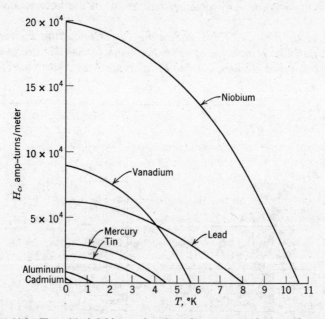

Figure 11.2   The critical field as a function of temperature for several superconducting elements. Niobium and vanadium are Type II superconductors and the rest are Type I.

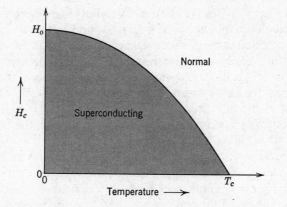

Figure 11.3   The critical field versus temperature for a superconductor.

to restore the normal resistivity is called the critical field, H_c. Likewise, the superconductivity vanishes if the critical current density J_c is exceeded. Both H_c and J_c depend on the temperature, and on each other. Figure 11.2 shows the critical field as a function of temperature, at zero current for seven elements. The curves are nearly parabolic and can be reasonably well represented by the relation:

$$H_c = H_0 \left(1 - \frac{T^2}{T_c^2}\right) \tag{11.1}$$

where H_0 is the critical field at $0°$K. Equation 11.1 is really the equation of a phase boundary, as Figure 11.3 shows. The critical current density is not a simple function of H and T; J_c vanishes at H_c and T_c, and is expected to approach zero gradually near H_c and T_c.

11.3 SUPERCONDUCTIVITY AND THE PERIODIC TABLE

Table 11.1 indicates certain regularities in the occurence of superconductivity in the pure elements. The Group V and Group VII metals tend to have high transition temperatures, and the even numbered groups, low transition temperatures. Alloys and intermetallic compounds follow the same general rule, provided an atom fraction average of the column numbers is taken. A 50%

Mo (Group VI)-50% Ti (Group IV) alloy for example, has an average column number of V, and, has a higher T_c than either Mo or Ti. As functions of this average, the critical temperatures of alloys and compounds also tend to vary periodically like the pure elements. The highest value of T_c does not occur exactly at a valence electron-to-atom ratio of five, but at 4.75. Thus, an alloy of 75 atomic percent Nb (Group V) and 25 atomic percent Zr (Group IV) has the optimum ratio, and, in fact, the highest T_c of any Nb-Zr alloy.

Figure 11.4 summarizes the observations of Matthias (1957) and other investigators on the periodic behavior of T_c. Table 11.2 lists T_c and H_c for a number of high critical field alloys and compounds. Many of the compounds have, at room temperature, the so-called beta-tungsten structure, A_3B, which is described in Figure 11.5. The B atoms form a BCC lattice; the A atoms are located in tetrahedral sites in three orthogonal chains. There is reason to believe that the chains of A atoms are connected with the

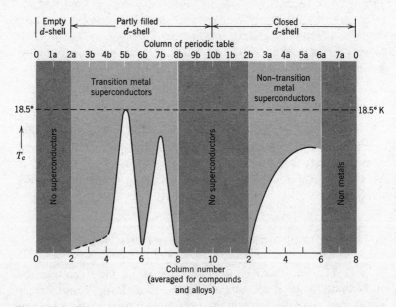

Figure 11.4   The variation of T_c with position in the periodic table.   (From B. T. Matthias, *Progress in Low Temperature Physics*, Vol. II, p. 138, ed. by C. J. Gorter, North Holland Publishing Co., Amsterdam, 1957.)

MATERIAL	STRUCTURE	T_c (°K)	H_c AT 4.2°K (AMP-TURNS/METER)
Nb$_3$Sn	Beta-tungsten	18.5	1.6×10^7 (Approx.)
Nb$_3$Al	Beta-tungsten	18.0	(?)
V$_3$Si	Beta-tungsten	17.0	1.3×10^7 (Approx.)
V$_3$Ga	Beta-tungsten	16.8	2.8×10^7 (Extrapolated)
NbN	FCC (rock salt)	16	0.8×10^7 (Approx.)
Nb-25%Zr	BCC solid solution	10.8	0.56×10^7
Nb-60%Ti	BCC solid solution	8.7	0.8×10^7

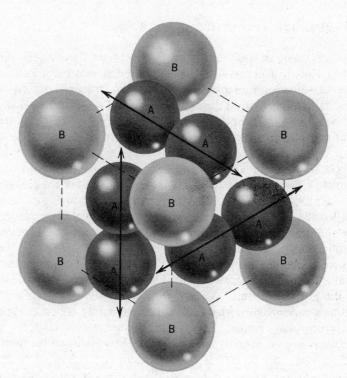

Figure 11.5 The beta tungsten structure. A atoms sit in tetrahedral sites, on faces of a BCC cell of B atoms. The arrows indicate continuation of chains of A atoms.

high T_c and H_c, as anything which breaks the chains leads to sharp drops in both. Recent evidence indicates that some of the beta-tungsten superconductors actually transform to other structures at liquid helium or hydrogen temperatures, by martensitic transformations. In any event, the beta-tungsten superconductors generally have the highest H_c and T_c.

There are other empirical rules for the occurrence of superconductivity. No superconductors have average column numbers of less than two. There are no superconductors which are insulators above T_c, few which are semiconductors and superconductivity is rare or nonexistent (the issue is still in question) in ferromagnetic and antiferromagnetic materials.

11.4   MAGNETIC PROPERTIES OF SUPERCONDUCTORS

The magnetic properties of superconductors are as remarkable as their electrical properties. Let us note first that the magnetic properties of superconductors are sensitive to structural defects of all kinds. In this section the behavior of ideal structurally perfect materials, free of imperfections is discussed. The structure sensitivity of all the properties is discussed in Sections 11.8 and 11.9. The ideal magnetic behavior of superconductors falls into two classes: Type I and Type II. Type I superconductors are completely diamagnetic; that is, $B = 0$ inside in the interior or, equivalently, $M = -H$. This behavior, illustrated in Figure 11.6, is called the *Meissner effect:* all of the magnetic field is excluded from the superconductor. Figure 11.6 also illustrates a geometric complication: when the specimen becomes superconducting, the field is concentrated at the sides of the specimen. It is, therefore, possible to exceed the critical field at the sides of the specimen, but not the top or bottom. When this happens, the specimen must exist as a mixture of normal and superconducting regions, called the *intermediate state*. The intermediate state is completely analogous to any mixture of two phases. Discussion of the intermediate state can be avoided if the description is limited to the magnetization of long cylinders parallel to the applied field. The magnetic field is uniform over the surfaces of such specimens. Consequently, they are either superconducting (fields less than

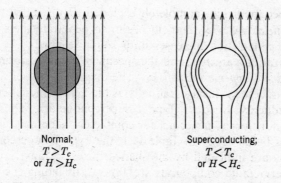

Normal;
$T > T_c$
or $H > H_c$

Superconducting;
$T < T_c$
or $H < H_c$

Figure 11.6   The Meissner effect.   When the specimen becomes superconducting, all of the magnetic field is expelled from it.   Ideal Type I superconductors behave in this manner, as do ideal Type II superconductors in magnetic fields below H_{c1}.

H_c) or normal (fields more than H_c), but never exist in the inter-mediate state.   A long cylinder of Type I material, parallel to the field, will exclude the field completely if it is below H_c, and be completely penetrated by magnetic flux if the field is above H_c. The magnetization of the normal state is usually negligible com-pared to the magnetization of the superconducting state and can therefore be neglected.

Type II superconductors behave differently, as Figure 11.7 shows.   For applied fields below H_{c1} the material is diamagnetic

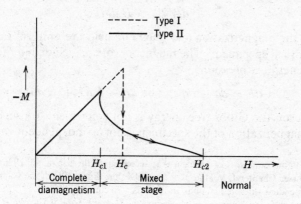

Figure 11.7   Magnetization curves for a Type I and Type II superconductor, having the same free-energy difference between the normal and superconducting states. The area under both magnetization curves is the same.

and hence the field is completely excluded. H_{c1} is called the *lower critical field*. At H_{c1} the field begins to penetrate the specimen, and the penetration increases until H_{c2} is reached. At H_{c2}, the magnetization vanishes, and the specimen becomes normal. H_{c2} is called the *upper critical field*. The magnetization of a Type II superconductor vanishes gradually as the field is increased, rather than suddenly as for a Type I superconductor. The Type II superconductor is, nevertheless, completely superconducting for all fields below H_{c2}. In Table 11.1, the Type II superconducting elements are indicated by the shaded boxes. Many more alloys and intermetallic compounds are Type II, including the high H_c, high T_c materials listed in Table 11.2.

The magnetization curves in Figure 11.7 are reversible for ideal superconductors whether Type I or Type II. The magnetization at any point on the magnetization curves of Figure 11.7 is therefore independent of the field and temperature paths followed in attaining the point. Non-ideal superconductors, in contrast, exhibit irreversible magnetization behavior.

11.5   THERMODYNAMICS OF SUPERCONDUCTORS

The magnetization of a solid involves the expenditure of energy. The work done per unit volume is given by

$$dW_{\mathrm{mag}} = H \, dM \tag{11.2}$$

since the magnetization and magnetic field are uniform in a long cylindrical specimen. The magnetic work is included in the Gibbs free energy expression:

$$dF = dq - dq_{\mathrm{rev}} + V \, dP - S \, dT + H \, dM \tag{11.3}$$

However, the Gibbs free energy is not convenient to use, because the magnetization of the specimen is not an independent variable.*

---

* This is the same sort of problem that accompanies the use of the Helmholtz free energy (see Chapter 1, Volume II). It involves the differential of the volume, which is a dependent variable for solids and liquids. To eliminate the term $(-P \, dV)$ in the differential of the Helmholtz free energy, the term PV is added to the free energy. The resulting function is the Gibbs free energy.

Let us modify in this case the Gibbs free energy by subtracting HM, to eliminate the term $H\,dM$:

$$dG = d(F - HM) = dq - dq_{\text{rev.}} + V\,dP - S\,dT - M\,dH$$

(11.4)

where G is called the *Gibbs magnetic function.*

At equilibrium for constant pressure, temperature and magnetic field

$$dG]_{\text{P,T,H}} \leq 0 \qquad (11.5)$$

At constant pressure and temperature, the free energy G_S of the superconductor as a function of magnetic field is obtained by intergrating the $-M\,dH$ term of Equation 11.4:

$$G_S(H) - G_S(0) = -\int_0^H M\,dH \qquad (11.6)$$

The normal state of most superconductors is paramagnetic and the magnetization is small compared with that of the superconducting state shown in Figure 11.7 for fields below H_c or H_{c2}. Therefore, neglecting the normal state magnetization

$$G_N(H) - G_N(0) = 0 \qquad (11.7)$$

where G_N is the free energy of the normal state.* At the critical field, the free energies of the normal and superconducting state must be equal for two-phase equilibrium:

$$G_N(H_c) = G_S(H_c) = G_S(0) - \int_0^{H_c} M\,dH \qquad (11.8)$$

from Equation 11.6. Combining Equations 11.8 and 11.7, the free energy difference ΔG between the two states becomes

$$\Delta G = G_N(0) - G_S(0) = -\int_0^{H_c} M\,dH \qquad (11.9)$$

The free-energy difference is just the area under the magnetization $(-M, H)$ curve. The Type I and Type II magnetization curves

---

* For the high-field superconductors such as those in Table 11.2, the assumption of Equation 11.7 is not proper.

are drawn in Figure 11.7 to give the same free energy difference; the areas under the curve are equal. For a Type I superconductor,

$$\Delta G = \int_0^{H_c} M dH = \tfrac{1}{2}H_c^2 \tag{11.10}$$

The term *thermodynamic critical field* H_c of a Type II superconductor is defined by setting the right-hand side of Equation 11.10 equal to the area under the Type II magnetization curve.

Since the free energy difference at zero field can be obtained from specific heat measurements, Equations 11.9 and 11.10 can be used to calculate the critical fields from specific heat data. Such calculations are accurate only for ideal or nearly ideal superconductors.

11.6   PENETRATION DEPTH AND THIN FILMS

In 1935, F. London and H. London described the Meissner effect and zero resistivity by adding two new equations to the four Maxwell equations for electricity and magnetism. According to the London equations, the applied field does not suddenly drop to zero at the surface of a Type I superconductor. Instead, H decays exponentially according to

$$H = H_0 e^{-x/\lambda} \tag{11.11}$$

where H_0 is the applied field at the surface and x is the distance in from the surface. As a consequence, the field is fairly large over a distance λ from the surface. The *penetration depth,* λ ranges from 300 to 5,000 A.U., depending on the material. Figure 11.8 shows the penetration of the magnetic field into a bulk specimen and into a thin film whose thickness is less than the penetration depth. In ordinary specimens whose dimensions are much larger than 5,000 Å, the major fraction of the volume is not penetrated by the field, and hence $M \cong -H$. A thin film does not show the same diamagnetism as a large specimen since it is at least partially penetrated by the applied field. As described in the previous section, the specimen becomes normal when the free energy of the superconducting state is raised by its diamagnetism to that of the normal state. The free energy necessary to bring about the

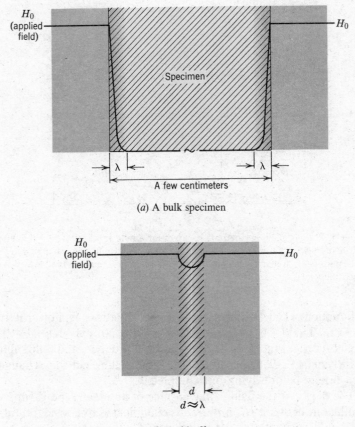

H_0 (applied field)

H_0

Specimen

λ λ

A few centimeters

(a) A bulk specimen

H_0 (applied field)

H_0

d

$d \approx \lambda$

(b) A thin film

Figure 11.8   Field penetration in a bulk and in a thin film, superconductor.

transition is just the area under the $-M$ versus H curve. The thin film, since it is penetrated by the field, has a much shallower $-M$ versus H curve. The area under the curve must be the same as that of the bulk material since we are dealing with the same phases, and therefore the same free energy difference. Therefore, the $-M$ versus H for a thin film continues out to higher fields, as shown in Figure 11.9. The thinner the film, the higher is the critical field. The H_c of experimental Type I films is found to be 10 to 20 times greater than that of bulk specimens.

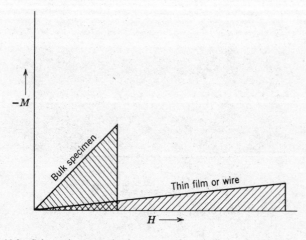

Figure 11.9 Schematic magnetization curves for a bulk specimen, and a thin film or wire of thickness, approximately, λ. Both are cylindrical in shape and parallel to the applied field, so that little distortion of the applied field occurs. Both are Type I superconductors.

Equation 11.11, while a good approximation, is not strictly correct. The decay of the magnetic field is not exponential with penetration. According to the Bardeen, Cooper, and Schrieffer (BCS) theory (1958) of superconductivity, the field may change sign briefly before dying out completely.

The decay of the field at the surface of an ideal Type II superconductor, between H_{c1} and H_{c2}, is complicated by considerations that are discussed later. It is sufficient for the present to remember that the field penetrates Type I superconductors and Type II superconductors if H is below H_{c1} to a small but definite extent. The field decays within a distance λ of the surface, where λ may be between a few hundred and a few thousand angstroms, depending upon the superconductor.

11.7   SURFACE ENERGY

In order to explain the Meissner effect in Type I superconductors it is necessary to postulate a positive surface energy between the normal and superconducting phases. If there were no surface energy between the normal and superconducting phases, the magnetic free energy of the sample could be lowered by having alternate layers of normal and superconducting phases, as shown in Figure 11.10. The normal layers would be very thin and small in volume, and the superconducting layers would have a thickness of the order of the penetration depth λ. The material would then be almost completely penetrated by the magnetic field, and the magnetic free energy lowered considerably. The specimen would still be essentially all superconducting. To explain why such behavior does *not* occur, it is necessary to assume the existence of a positive surface energy between the phases, which is large enough to offset the potential reduction in magnetic energy.

Now let us consider the possibility of a negative surface energy in the two phase structure. In this case, the field will readily penetrate the specimen and the free energy of magnetization will be lowered. The superconducting state should then persist to higher magnetic fields, as the result of the field penetration. In 1957, Abrikosov calculated the magnetic properties for a negative surface energy superconductor, predicting magnetization curves as depicted in Figure 11.7 for the Type II superconductor. Ac-

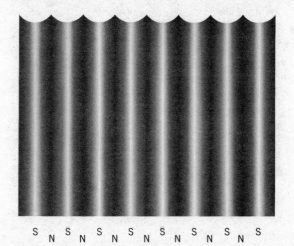

S   S   S   S   S   S   S
  N   N   N   N   N   N

Figure 11.10 A schematic of alternating layers of superconducting and normal phases in a hypothetical superconductor. The thickness of superconducting layers is of the order of λ; that of the normal layers, much less. The solid, bulk specimen is almost completely penetrated, lowering the free energy of magnetization. A positive surface energy between the phases would prevent the situation illustrated.

cording to the Abrikosov theory, the magnetic field penetrates the specimen in lines rather than layers. Figure 11.11 shows one possible configuration. The normal regions at the centers of the flux lines are surrounded by vortices or whirlpools of super-currents. A flux line, together with its current vortex, is called a *fluxoid.* At H_{c1}, fluxoids appear in the specimen, and increase in number as the magnetic field is raised. At H_{c2}, the fluxoids completely fill the crossection of the specimen and Type II super-conductivity disappears.

It is possible to predict whether a given superconductor will be Type I or Type II from the relation:

$$H_{c2} = \sqrt{2}\mathfrak{K}H_c \tag{11.12}$$

If the parameter $\mathfrak{K}$ is more than $1/\sqrt{2}$, and H_{c2} is more than H_c,

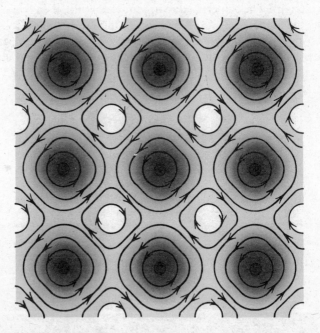

Figure 11.11   The penetration of the magnetic field into a negative surface energy superconductor. The shading indicates the field intensity, and the lines, the flow of current. The flux lines (which are normal to the paper and centered in the dark spots), surrounded by the current rings, are called fluxoids. (After A. A. Abrikosov, *J. Phys. Chem. of Solids,* Vol. II, p. 199, 1957.)

the material is a Type II superconductor; if κ is less than $1/\sqrt{2}$, it is a Type I superconductor. If the residual resistivity of the material in its normal state is small, the quantum theory of superconductivity gives

$$\mathcal{K} = \mathcal{K}_0 \cong 10^5 \left| \frac{dH_c}{dT} \right|_{T=T_c} T_c \lambda^2(0) \qquad (11.13)$$

where $\lambda(0)$ is the penetration depth for very weak fields. (λ is slightly sensitive to H.) If the normal state of the material has a very large residual resistivity,

$$\mathcal{K} = \mathcal{K}_l = 367\rho\left(\frac{\gamma}{V_m}\right)^{1/2} \qquad (11.14)$$

where ρ is the residual resistivity, V_m the molar volume, and γ the temperature coefficient of the electronic specific heat. For moderate residual resistivities, the empirical relation

$$\mathcal{K} = \mathcal{K}_0 + \mathcal{K}_l* \qquad (11.15)$$

holds. As ρ increases, $\mathcal{K}_l$ dominates.

Type II superconductors are important because they have high critical fields. All of the high field superconductors listed in Table 11.2 are Type II. All but very dilute alloys are Type II. Alloying raises the residual resistivity and, according to Equation 11.14,

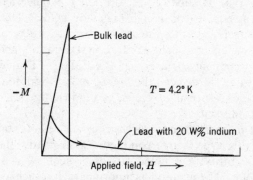

Figure 11.12   The effect of an alloy addition to a Type I superconductor. The addition of indium to lead (type I) has raised ρ, and therefore κ, over the critical value, so that the alloy became Type II. (After J. D. Livingston, unpublished.)

* κ_0 denotes the ideal contribution and κ_l that of imperfections.

raises $\mathcal{K}$. An example of the effect of alloying is shown in Figure 11.12. Pure lead is Type I, as the magnetization curve shows. If indium is dissolved in the lead, $\mathcal{K}$ increases until it is more than $1/\sqrt{2}$; the lead-indium alloy then becomes a Type II superconductor. Figure 11.12 indicates the Type II behavior of a Pb-20% In alloy.

Thin films of Type II superconductors do not behave in the manner outlined in Section 11.6 for Type I films. Even when a Type II film is only a penetration depth thickness the fluxoids move in and out as usual, and the bulk Type II properties persist.

11.8   HARD AND SOFT SUPERCONDUCTORS

Supercurrents will flow in a superconductor only where the magnetic field has penetrated. Ideal Type I superconductors which carry currents only on the surface, are therefore relatively poor current carriers. Although ideal Type II superconductors can be thoroughly penetrated by the field, they also have a small J_c. However, this is not the case, as the fluxoids are mobile. When a current flows in the material the fluxoids experience a force proportional to $\mathbf{J} \times \mathbf{B}$ (the Lorentz Force) which drives them out of the superconductor and the material becomes normal.

Non-ideal Type II superconductors, on the other hand, carry large supercurrents, even at high magnetic fields. Their critical current density is extremely sensitive to microstructure. Figure 11.13 shows the effect of cold work, recrystallization, and precipitation heat treatment on a Nb-25% Zr alloy. The higher the density of structural defects, the higher is J_c. Addition of interstitial impurities, especially oxygen, can induce fine scale precipitation which raises J_c.

The exact mechanism by which structural defects stabilize the current-carrying Type II superconducting state against the Lorentz force is still controversial. One proposed mechanism is pictured in Figure 11.14. The fluxoid has a normal region such as an oxide or precipitate particle at its center. If the center of the fluxoid and the particle are superimposed, the two normal regions become one. The volume of superconducting material thereby increases at no cost in magnetic energy, and the free energy is reduced. The fluxoid must increase the free energy in order to move away from

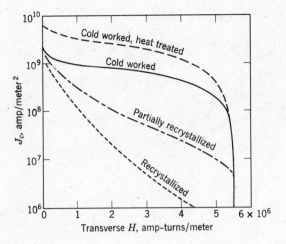

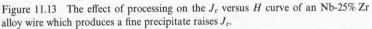

Figure 11.13   The effect of processing on the J_c versus H curve of an Nb-25% Zr alloy wire which produces a fine precipitate raises J_c.

the particle. It is thus *pinned* to the imperfection by the energy requirement. An alternative theory of fluxoid pinning involves heterogeneous nucleation of fluxoids at imperfections or precipitates. It has been used to account for the high critical fields of

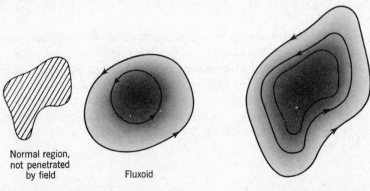

Normal region, not penetrated by field

Fluxoid

(a) Fluxoid is attracted to normal region which is not penetrated by the field

(b) Fluxoid sits in normal region, lowering free energy of the super conductor

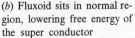

Figure 11.14   Fluxoid pinning by a normal, unpenetrated inclusion. A proposed mechanism for the high J_c of nonideal Type II superconductors.

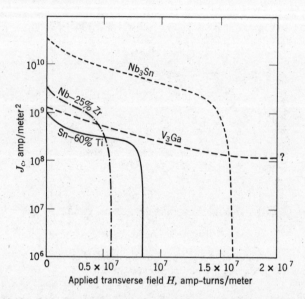

Figure 11.15   Critical current density versus applied field for superconducting wire for high-field application, at 4.2°K. The critical field of V_3Ga at 4.2°K has not been measured but is thought to be higher than Nb_3Sn.

some of the superconductors listed in Table 11.2 and Figure 11.15. Whatever mechanism is responsible the important technological observation is that structural imperfections lead to high critical current density.

Since fluxoid pinning in nonideal superconductors leads to hysteresis and irreversibility in the magnetization curve, such materials are often called *hard superconductors*. A normal structural imperfection in either Type I or Type II superconductors is surrounded by a supercurrent circuit which can freeze magnetic flux in or out (Problem 11.10). The pinned fluxoid structure is more important in Type II superconductors for it allows supercurrent to flow at high fields.

11.9   SUPERCONDUCTING MAGNETS

Of the many possible applications of superconducting materials reported since 1911, the high field solenoid has become the most useful. Commercial production of solenoids capable of exceeding

~2.2 × 10⁶ amp/m (~30,000 Kgauss) followed the development of nonideal Type II Nb$_3$Sn and Nb-Zr wire by Kunzler and co-workers (1961) at the Bell Laboratories. Some of their small solenoids are shown in Figure 11.16.

The cost of a conventional solenoid installation capable of producing a magnetic field of 1 × 10⁷ amp/m (~100 Kgauss) in a volume of ~50 cc is more than ten times that of a superconducting solenoid wound with Nb$_3$Sn wire. Such a conventional solenoid furthermore requires 2000 Kw of power and 1000 gal/min of cooling water as well as considerable space for auxiliary equipment. A superconducting solenoid, in contrast, occupies little space, has no steady-state power consumption and uses relatively little liquid helium coolant. Although their use is still largely confined to laboratory research, superconducting solenoids may some day prove useful in magnetohydrodynamic power conversion and in other engineering areas.

If the a-c properties of the high field, high current superconductors were as good as their d-c properties, many useful devices could be built. In an alternating field, the magnetic hysteresis which accompanies high J_c leads to local hot spots which drive the superconductor normal. Some progress has nevertheless been

Figure 11.16 First effective superconducting magnetic solenoids made by Bell Laboratories using Mo-Re (smallest), Nb$_3$Sn (largest) or Nb-Zr wire (dark). A field of 56 × 10⁵ amp/meter (70 K-oersteds) was obtained at 4.2°K with the largest.

made using pure niobium instead of alloys. Suffice to say transformers, motors, generators and similar electrical devices made of niobium have not reached the production stage.

DEFINITIONS

Normal State. A superconducting solid in the normal state exhibits measurable electrical resistance.

Superconducting State. A solid in the superconducting state exhibits zero electrical resistance.

Critical Temperature (T_c). The temperature above which a superconductor regains its normal electrical resistance.

Critical Current Density (J_c). The current density above which superconductivity disappears.

Critical Field H_c. The magnetic field above which superconductivity disappears.

Meissner Effect. The expulsion of magnetic flux ($B = 0$) in a superconductor.

Penetration Depth λ. The effective depths to which the magnetic field penetrates a superconductor.

Soft Superconductor. A substance which exhibits reversible ideal magnetization and low J_c, whether Type I or Type II.

Hard Superconductor. A substance characterized by irreversible magnetization and a high value of J_c, whether Type I or Type II.

Type I Superconductor. One exhibiting essentially complete flux expulsion, due to a positive surface energy between the normal and superconducting phases.

Type II Superconductor. One in which magnetic flux partially penetrates, due to a negative surface energy existing between the normal and superconducting phases. A Type II superconductor remains superconducting at applied fields above the thermodynamic critical field.

Thermodynamic Critical Field (H_c). The critical field calculated, assuming complete flux expulsion takes place. H_c may be calculated from specific heat measurements or from the area under the magnetization curve.

Lower Critical Field (H_{c1}). The field at which magnetic flux first penetrates an ideal Type II superconductor.

Upper Critical Field (H_{c2}). The field at which superconductivity vanishes for an ideal Type II superconductor.

Fluxoid or Flux Thread. A microscopic normal region surrounded by circulating supercurrents in a Type II superconductor at fields between H_{c1} and H_{c2}.

Mixed State.   The fine subdivision of normal and superconducting phases in a Type II superconductor that arises between H_{c1} and H_{c2}; the fluxoid or flux thread configuration.

Fluxoid Pinning.   The attractions of fluxoids by lattice defects, which tend to immobilize fluxoids.

Intermediate States.   A mixture of normal and superconducting regions, which occurs due to inhomogeneous magnetic fields at the surface of the specimen.

BIBLIOGRAPHY

F. London, *Superfluids,* Volume I, Dover, New York, 1961.

E. A. Lynton, *Superconductivity,* 2nd Ed., Methuen-Wiley, New York, 1964.

B. T. Matthias, "Superconductivity in the Periodic System," p. 138 of Volume II, *Progress in Low Temperature Physics,* C. J. Gorter, Ed.,

V. L. Newhouse, *Applied Superconductivity,* Wiley, New York, 1964.

D. Shoenberg, *Superconductivity,* Cambridge University Press, 2nd Edition,

PROBLEMS

11.1   Describe, with sketches if necessary, at least two ways to measure the transition temperature of a superconductor.   (For references, if you wish, try: White, G. K., *Experimental Techniques in Low Temperature Physics,* Oxford U. Press, 1959; Hoare, F. E., Jackson, L. C., and Kurti, N., *Experimental Cryophysics,* Butterworths, London, 1961; and Jackson, L. C., *Low Temperature Physics,* Methuen-Wiley, New York, 5th Edition 1962.)

11.2   A specimen of V_3Ga has a transition temperature of $14.5°K$. At $14°$, the critical field is 1.4×10^5 amp-turns/meter, and at $13°$, 4.2×10^5 amp-turns/meter.   Use Equation 11.1 to extrapolate the critical field back to $4.2°$ and $0°$. (Differentiate Equation 11.1, and match the slope of the data above to the result.)   Compare the results with Table 11.2.

11.3   Derive the relation:

$$C_s - C_n = V_m T \left[\frac{dH_c}{dT} \right]_{T=T_c}^2$$

for a Type I superconductor, where C_s and C_n are the molar heat capacities for the superconducting and normal states, and V_m the molar volume. *Hint.*   Apply the relations:

$$\frac{\partial \Delta F}{\partial T} = \Delta S \quad \text{and} \quad \frac{\partial \Delta S}{\partial T} = \frac{\Delta C_P}{T}$$

11.4   Use Nordheim's law (see Chapter 4) to show that, in dilute alloys, H_{c2} is proportional to the concentration of solute or impurity element.

11.5   For a given Type II superconductor, H_{c1} must decrease as H_{c2} increases. Why?

11.6   The thermodynamic critical field of lead is 65,000 amp-turns/meter; since lead is a Type I superconductor, this is also the bulk critical field of a soft specimen. A thin film of lead, whose thickness is not known, is found to have a magnetization $M = -H/20$. Estimate the critical field of the lead film. What can you say about the thickness of the film?

11.7   What experimental data will distinguish a Type I superconductor from a Type II superconductor? If you had to make such a distinction, what experiments would you make?

11.8   When Type II alloys are age-hardened, the critical current density rises, but H_{c2} decreases. Explain how this could occur.

11.9   Using a ruler or scale, determine $\mathcal{K}$ for the Pb-20% In alloy in Figure 11.12.

11.10   Use Faraday's law of induction to show that a hole (e.g., an impurity or imperfection) in a superconductor will freeze flux, that is, $dB/dt = 0$ in the hole. Remember that $E = 0$ for any circuit through the superconductor which encloses the hole, and also that the Meissner effect does not apply to the hole.

11.11   Refer to Figure 11.10. Suppose that the superconducting layers all had the thickness λ. Neglect the thickness of the normal layers. Assume that the specimen is then completely penetrated by the field. Show that, for a Type I superconductor, the surface energy γ must obey the condition:

$$\gamma > \frac{H_c{}^2\lambda}{4}$$

(*Hint.* Take a free-energy balance for a unit volume of the material, surface energy versus magnetic energy.)

11.12   In the Nb-Ti alloy system, at which compositions would you look for the highest T_c? Would you expect the highest critical fields to be found at these compositions, too? Why, or why not?

CHAPTER TWELVE

Dielectrics

Dielectrics are used in capacitors and as electrical insulation. The dielectric constant or relative permittivity of a material can vary with temperature and frequency, the bonding, crystal structure, phase constitution, and structural defects of the dielectric. All these factors influence the response of the induced or permanent electric dipoles in the dielectric to a steady or alternating electric field. If the polarization lags the applied field strength, it leads to an electrical energy loss which appears as heat and is proportional to the product of the relative permittivity and the tangent of the lag angle δ. Another undesirable energy loss in dielectrics arises from ion and electron migration. Overheating or cyclic heating leads to degradation of the dielectric and breakdown. The electric field strength at which an insulator breaks down is called its dielectric strength. Most dielectrics are, therefore, rated by three factors: (1) relative permittivity, (2) tangent of lag angle and (3) dielectric strength. Other dielectric properties, such as piezo-electricity and ferroelectricity, are also described in this chapter.

12.1 INTRODUCTION

The relative permittivity ϵ_r of a dielectric is defined as the ratio of the permittivity ϵ of the dielectric to the permittivity ϵ_0 of empty space

$$\epsilon_r \equiv \frac{\epsilon}{\epsilon_0} \qquad (12.1)$$

If a dielectric of relative permittivity ϵ_r were inserted into a parallel plate capacitor whose plate area A was very large, the capacity, for a plate spacing d,

$$C_0 = \frac{A}{d}\epsilon_0 \qquad (12.2)$$

would increase to

$$C = \frac{A}{d}\epsilon = \frac{A}{d}\epsilon_0\epsilon_r = C_0\epsilon_r \qquad (12.3)$$

That is, the dielectric has increased the capacity by a factor of ϵ_r. Figure 12.1 shows the capacitor plates, with the space between filled by the dielectric.

From the definition of capacity and current, we can write

$$V = \frac{Q}{C} = \int \frac{I\,dt}{C} \qquad (12.4)$$

and differentiate this relation, obtaining

$$I = C\frac{dV}{dt} \qquad (12.5)$$

for the current through the capacitor. If the voltage used is sinusoidal, that is, it varies as

$$V = V_0 \sin \omega t \qquad (12.6)$$

V_0 is the maximum value of the voltage and $\omega = 2\pi f$, where f is the frequency, and t is the time. Then

$$I = CV_0\omega \cos \omega t \qquad (12.7)$$

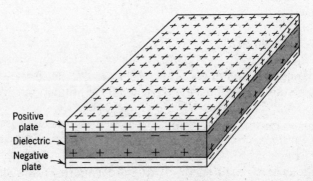

Figure 12.1   A parallel plate capacitor with dielectric between the plates. Free charge is shown on the plates, and induced charges on the surfaces of the dielectric.

Hence the current leads the voltage by 90°; or out of phase by a quarter-cycle. This applies only to the ideal case, i.e. when a loss-free dielectric is in the capacitor. When a real dielectric is present, the current leads the voltage by 90-δ, where the angle δ is a measure of the dielectric power loss.

$$\text{Power Loss} = \pi f V_0^2 \, \epsilon_r \tan\delta \qquad (12.8)$$

Such a phase retardation of the current through the capacitor is exactly analogous to the case of anelasticity that we covered in Volume III (Section 3.5), where the strain lagged the stress. The product $\epsilon_r \tan\delta$ is called the loss factor and $\tan\delta$ the loss tangent or dissipation factor. The loss factor consequently characterizes the usefulness of a material as a dielectric or as an insulator; in both cases a low loss tangent is desirable.

The third factor of major significance in judging the usefulness of an insulator is called the dielectric strength or breakdown strength. It is defined as the maximum voltage gradient which a dielectric will withstand before failure occurs. Its experimental value is affected by the geometry of the specimen, the electrodes and other aspects of testing procedure. A large scatter of test results is usually found. Such data can, nevertheless, serve as a guide in judging and developing dielectric materials. Dielectric breakdown begins with the appearance of a number of electrons in the conduction band; these electrons are accelerated rapidly by the high field in the dielectric, and attain high kinetic energies. Some of the kinetic energy is transferred, by collisions, to valence electrons, which are thereby elevated to the conduction band. If a large enough number of electrons initiate this process, it multiplies itself, and an avalanche of electrons is loosed in the conduction band. The current through the dielectric increases rapidly, and the dielectric is apt to locally melt, burn, or vaporize. Conduction of electrons are required to initiate the process. These may originate in a number of ways. A common origin is arcing between a high potential lead and the contaminated surface of an insulator. Impurity atoms can also donate electrons to the conduction band. Interconnecting pores in dielectrics sometimes provide direct breakdown channels as a result of electrical gas discharge. Where a dielectric is subject to high field over a long period, breakdown is generally preceded by local melting. In old

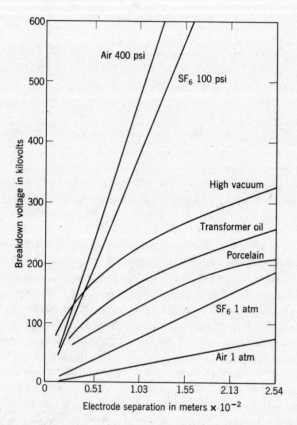

Figure 12.2   D-c breakdown or dielectric strength of various solids, liquids, gases, and vacuum, in uniform fields.   Breakdown voltage versus dielectric thickness is plotted. (From J. Trump, in A. von Hippel, *Dielectric Materials and Applications,* Wiley, New York, 1954.)

capacitors breakdown can occur at relatively low field strengths if the dielectic has suffered chemical and mechanical abuse. Figure 12.2 shows the breakdown voltages for various insulating media.

12.2   FIELD VECTORS

According to a theorem of Gauss, the total flux ϕ emanating from a closed surface is equal to the charges enclosed by that surface or

$$\phi = \epsilon Q = \int^s \mathbf{D} \cdot d\mathbf{S} \tag{12.9}$$

where the vector $d\mathbf{S}$ has the magnitude dS and is normal to the surface elements it represents. The magnitude of the flux density D within a dielectric of relative permittivity ϵ_r located between the plates of a large parallel plate condenser whose surface charge density (see Figure 12.1) is equal to σ, is simply σ. The field strength between the plates is therefore

$$\mathcal{E} = \frac{D}{\epsilon_0 \epsilon_r} = \frac{\sigma}{\epsilon_0 \epsilon_r} \tag{12.10}$$

When a dielectric is inserted between the charged plates of a capacitor, electric dipoles are formed whose electric moment $\mathbf{p}_e$ is given by

$$\mathbf{p}_e = q\mathbf{r} \tag{12.11}$$

where q is a positive charge separated from a negative charge of the same size by the distance r. The origin of the vector is at the negative charge. The total dipole moment per unit volume is called the polarization $\mathbf{P}$. Using Equations 12.10 and 12.11,

$$\mathbf{P} = \epsilon_0 \mathcal{E}(\epsilon_r - 1) \tag{12.12}$$

provided only the polarization of the dielectric is considered and the *free-space* polarization is subtracted.

12.3   POLARIZATION

The field that a molecule in the interior of a dielectric situated between the plates of a charged condenser actually experiences is known to be larger than the applied field $\mathcal{E}$. This is related to the polarization which occurs within and on the surfaces of the dielectric. The actual field acting on the molecule is therefore called the local field $\mathcal{E}_{loc}$. The dipole moment induced in the molecule by the local field is given by

$$\mathbf{p}_{mol} = \alpha \mathcal{E}_{loc} \tag{12.13}$$

where $\mathbf{p}_{mol}$ represents its moment and α is called the *polarizability* of the molecule. For dielectrics containing N molecules per unit volume, the total dipole moment or polarization is

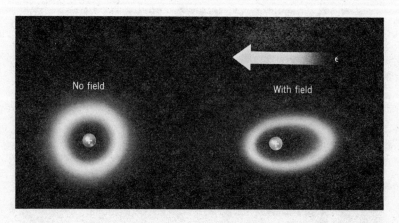

Electronic polarization: the charge cloud around the nucleus is distorted by the field

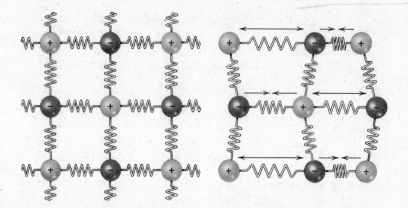

Ionic polarization: the field distorts the lattice

Orientation polarization

Figure 12.3 Schematic representation of different types of polarization applicable to solids, liquids and gases. (After A. R. Von Hippel.)

256

$$\mathbf{P} = N\alpha\mathcal{E}_{\text{loc}} \tag{12.14}$$

Substituting Equation 12.14 in Equation 12.12 gives,

$$\epsilon_r - 1 = \frac{\mathbf{P}}{\epsilon_0\mathcal{E}} = \frac{N\alpha\mathcal{E}_{\text{loc}}}{\epsilon_0\mathcal{E}} \tag{12.15}$$

Thus, the relative permittivity is related to P, α, $\mathcal{E}_{\text{loc}}$ and $\mathcal{E}$.

The four major types of polarization which occur in dielectrics are shown schematically in Figure 12.3. One or two of them are always present and at a particular frequency of applied field or temperature, one or another may play a major role. The top schematic represents electronic polarizability (α_e) which prevails in all dielectrics, since it is due to the displacement of the negative electron cloud of each atom from its nucleus. Ionic polarizability (α_i) is due to the displacement of adjacent ions of opposite sign and is only found in ionic substances. Orientation polarizability (α_o) occurs in liquids and solids which have asymmetric molecules whose permanent dipole moments can be aligned by the electric field, in the same manner as magnetic moments are aligned by a magnetic field. The randomizing effect of thermal energy is also there, and the orientation polarizability α_o is calculated in exactly the same manner as the paramagnetic susceptibility. (See Chapter 9.) The Curie law, therefore, also applies to α_o

$$\alpha_o = \frac{CP_o^2}{T} \tag{12.16}$$

Space charge polarization

Figure 12.3   (Continued)

where C is the Curie temperature, P_o the permanent dipole moment and T the absolute temperature. (At low temperatures dipole orientation by the field is opposed by frictional forces.) Although orientation polarization of the above kind is not found in ionic crystals and glass, another orientational mechanism is. It appears when two or more equivalent positions for an impurity ion adjacent to a vacancy are available in the substance. The ion-vacancy dipole can then change positions in the field, as shown in Figure 12.4.

The fourth type of polarizability shown in Figure 12.3 is due to the accumulation of charges at phase interfaces in multiphase

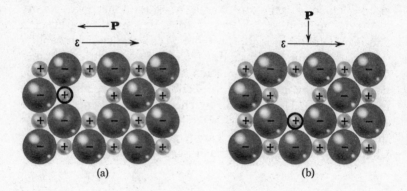

(a)   (b)

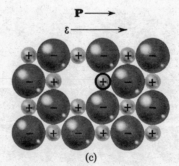

(c)

Figure 12.4   Schematic of ion jump polarization. A cation impurity of larger charge and an anion vacancy form a net dipole, and reorient when the field is applied.

dielectrics.   Such *space charge polarizability* is possible when one of the phases present has a much higher resistivity than the other. It is found in ferrites and semiconductors and in composite insulators at elevated temperatures.

12.4   FREQUENCY AND TEMPERATURE DEPENDENCE OF ϵ_r

The total polarization **P**, the total polarizability α, and the relative permittivity ϵ_r of a dielectric in an alternating field all depend on the ease with which the dipoles can reverse alignment with each reversal of the field.   As already indicated in Sections 12.1 and 12.3 some polarizability mechanisms do not permit sufficiently rapid reversal of the dipole alignment.   This situation is typical of the mechanically disturbed system met with in the discussion of anelasticity (see Sections 3.5 and 3.6 of Volume III).   In such a process the time required to reach the equilibrium orientation is called the relaxation time, and its reciprocal, the relaxation frequency.   When the frequency of the applied field exceeds that of the relaxation frequency of a particular polarization process, the dipoles cannot reorient fast enough and operation of the process ceases.   Since the relaxation frequencies of all four polarization processes differ, it is possible to separate the different contributions experimentally.   The result is shown in the upper diagram of Figure 12.5.   Four frequency ranges are found over which the separate polarization mechanisms operate.

In such a highly covalent solid as diamond, only electronic polarization is present.   Thus, ϵ_r can be measured optically from the index of refraction (see Chapter 13).   Ionic materials cease to contribute to the total polarization at infrared frequencies.   With such materials it is possible to measure separate ionic and electronic contributions by making both optical and electrical measurements of ϵ_r.   The orientation and space-charge polarization only function at lower frequencies.

The effect of temperature on the dielectric constant of ionic and electronic materials is, in general, small at low temperature, but increases with increasing temperature.   At elevated temperatures,

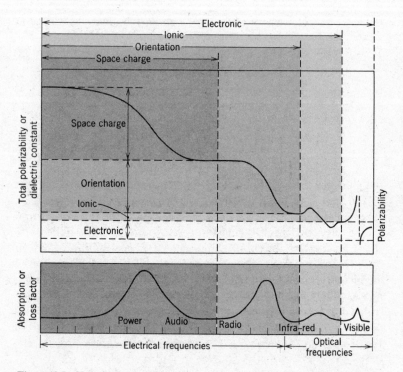

Figure 12.5   Variation of the total polarizability and dielectric absorption as a function of frequency.   Each contribution to the polarizability decays as its characteristic resonant frequency is exceeded.   (After E. J. Murphy and S. D. Morgan, *Bell System Tech. Il.,* **16**, 493, 1937.)

ion mobility is appreciable.   The dependence on temperature of orientation polarization can be large, as shown in Figure 12.6 for nitrobenzene (C_6H_6NO).

The combined effect of temperature and frequency is quite important in ionic materials, as shown in Figure 12.7.   It results in a sharp rise in ϵ_r in soda-lime glass at approximately 90°C and increases more rapidly with rising temperature the lower the frequency.   Similar observations have been made on alumina ceramics.   The increases may be attributed to an increase in both ion-jump orientation and space-charge polarization.   (The exponential increase in electrical conductivity with temperature may also contribute.)

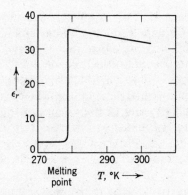

Figure 12.6 Relative permittivity of nitrobenzene as a function of temperature.
[Data of C. P. Smyth and C. S. Hitchcock, *J. Am. Chem. Soc.*, **55**, 1296 (1933).]

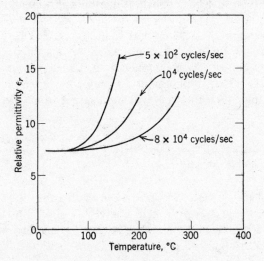

Figure 12.7 Effect of frequency and temperature on the permittivity of a soda-lime-silica glass. [Data of M. J. O. Strutt, *Arch. Elektrotech*, **25**, 715 (1931).]

12.5 ENERGY LOSS

The energy losses which occur in dielectrics are due to d-c conductivity and dipole relaxation. As mentioned previously, the loss factor ($\epsilon_r \tan \delta$) of a dielectric is a useful indication of the energy lost as heat. Its variation with frequency is shown in the lower diagram of Figure 12.5. At frequencies in the optical and infrared ranges the maxima illustrate optical and infrared absorption. The maximum dielectric loss for any particular type of polarization process occurs when its relaxation period is the same as the period of the applied field, that is, when a resonance occurs. Thus, the maxima in Figure 12.5 all occur near the frequency limit of each particular polarization. The loss then falls off to either side of the maximum. This is to be expected, for then the relaxation time is either large or small compared to the period of the applied field.

Dielectrics may be divided into low and high loss materials. Typical high loss materials are polar organic materials. Ceramic materials of high dielectric constant like barium titanate are also high loss materials. The major energy losses in ionic crystals and glasses occur at frequencies of less than 10^4 cps. They may be attributed to ion-jump relaxation. Losses due to ion-vibration and deformation are seldom significant at the frequencies used in electronic and power applications. Conduction losses, however, are appreciable; they increase with decreasing frequency at low frequencies.

12.6   FERROELECTRICS

Some dielectrics are called *ferroelectrics,* because they exhibit a polarization versus electric field curve very similar to the *B-H* curve of a ferromagnetic material. Such a *ferroelectric hysteresis loop* is shown in Figure 12.8, with the *remanent polarization P_r* and the *coercive field $\mathcal{E}_c$.*

Domain structure has been observed in ferroelectric barium titanate ($BaTiO_3$) using polarized light. The hysteresis loop can be explained in a manner paralleling that used in Chapter 9, even though the process of growth is somewhat different.

The spontaneous polarization in ferroelectrics vanishes at some critical temperature called the *ferroelectric Curie temperature.* Below the Curie temperature, D and P are not linear functions of $\mathcal{E}$, hence in describing ϵ_r, it is more convenient to consider the

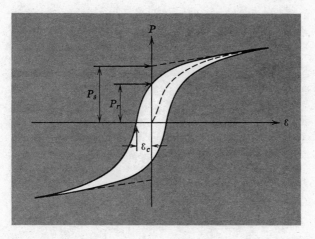

Figure 12.8   A ferroelectric hysteresis curve, with the coercive field $\mathcal{E}_c$, the saturation polarization P_s, and the remanent polarization P_r indicated.

initial relative permittivity.   Above the Curie temperature, the relative permittivity varies according to the Curie-Weiss law. Above the Curie temperature barium titanate ($BaTiO_3$) has the Perovskite structure, which is shown in Figure 12.9, in two different ways.   Figure 12.9*a* shows the Ti^{4+} ion at the center of a cubic unit cell, with oxygen ions at the faces and Ba ions at the corners. Figure 12.9*b* puts the Ba ion at the cube center, and Ti ions, octahedrally coordinated by oxygen ions, at the corners.   When a field is applied below the Curie temperature, the anions all move in one direction, and the cations in the other, destroying the cubic symmetry, and leaving the unit cell with a net dipole moment. When the field is reduced to zero the ferroelectric is left with a remnant polarization.   This can only be removed by applying a negative coercive field $\mathcal{E}_c$.

The relative permittivity of barium titanate also varies with temperature, as shown in Figure 12.10.   The two peaks shown at $5\,^{\circ}C$ and $-80\,^{\circ}C$ are also due to phase changes.   Such strong temperature dependence of the dielectric constant even over a small temperature range can be modified by solid-solution alloying, usually by additions of strontium titanate.   Such alloying also leads to a more linear response of D to $\mathcal{E}$.

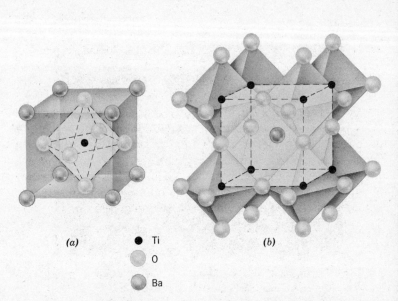

(a)   ● Ti
                       ○ O
                       ● Ba   (b)

Ion positions in ideal perovskite structure, i.e., BaTiO₃ above the Curie temperature

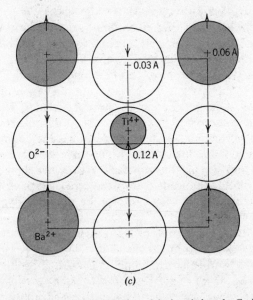

(c)

Top view of unit cell (a), showing the shifting of the ions below the Curie temperature

Figure 12.9   Crystal structure and ferroelectricity in barium titanate.

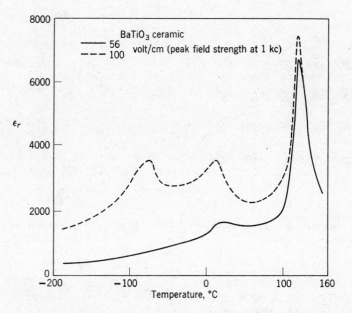

Figure 12.10   Permittivity of barium titanate ceramic as function of temperature. (Measurements of W. B. Westphal, Laboratory for Insulation Research, Massachusetts Institute of Technology.)

12.7   PIEZOELECTRICITY

Mechanical strains can polarize a crystal by displacing ions relative to one another provided that the crystal does not have a center of symmetry. By a center of symmetry, we mean a point in a crystal about which the lattice sites and atoms are symmetric. Figure 12.11 illustrates this condition; the center of symmetry cancels out all possible polarizations.

Piezoelectric crystals are used in devices called transducers which convert electrical to mechanical energy. These are used in microphones, phonograph pickups, strain gages, and sonar devices. Quartz, which is piezoelectric and not ferroelectric, is often used in such devices, but it has low sensitivity and requires voltage amplification. Rochelle salt ($KNaC_4H_4O_6$), a ferroelectric, is also used in transducer applications. It is readily attacked by moisture and can only be used in the temperature range $-18°C$ to $24°C$

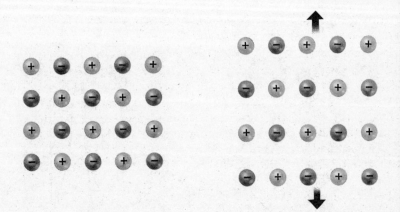

(a) Tensile strain of ionic crystal with center of symmetry (any lattice site).   No induced dipole moment

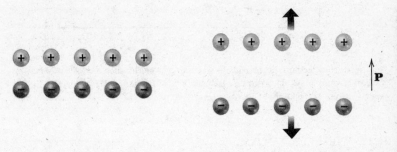

(b) Tensile strain of ionic crystal with no symmetry center, showing strain induced dipole moment

Figure 12.11   Piezoelectricity and crystal symmetry.   In crystals lacking a center of symmetry an elastic strain induces a dipole moment.

because it undergoes phase changes.   Barium titanate has less piezoelectric sensitivity than Rochelle salt, but it can be used over a wider range of temperature and is resistant to atmospheric attack.   Barium titanate piezoelectric materials are fabricated by ceramic processes in different shapes.   They are given a polarization treatment by cooling through the Curie temperature in a strong electrical field.   They can be used in this state up to temperatures of about 70°C.   For higher temperature use, lead titanate ceramics are employed.

12.8   USES OF DIELECTRICS

Useful dielectrics are divided into three principal categories: (1) those with an $\epsilon_r < 12$; (2) those with an $\epsilon_r > 12$; and (3) ferroelectrics and ferromagnetics. The first category may be broken down into subcategories based on operating temperature. Cotton, silk, paper, and a long list of polymers and liquids are used where the temperature does not exceed about 90°C. Inorganic fillers like mica and asbestos, bonded with organics, are only used up to about 130°C. Up to about 180°C, silicone binders are used with such fillers. Higher-temperature applications require insulation consisting entirely of mica, procelain, glass, and similar inorganic materials. The latter are also required in high-voltage applications to avoid breakdown. In such cases, it is important that the dielectric not absorb moisture so that a high dielectric strength prevails. Where the temperature is high as in furnaces, mechanical strength, as well as low resistivity, is important. In some high-frequency applications where mechanical strength and a low loss factor are required, only special ceramics such as forsterite (Mg_2SiO_4) are used.

Insulators with a relative permittivity between 12 and 100 have long been made of titania (TiO_2) base ceramics. Since Wainer and Solomon (1942) first showed that extremely high values of ϵ_r can be obtained with barium titanate, efforts to utilize this material have expanded. Its high-temperature sensitivity and low breakdown strength limit its electrical use to d-c or low frequency a-c applications.

DEFINITIONS

Field Strength or Intensity $\mathcal{E}$. The force experienced by a unit charge placed in an electric field.

Dielectric Displacement **D**. The vector used to describe the electric field in dielectrics. $D = \epsilon_0 \epsilon \mathcal{E}$; $D = \sigma$ on a capacitor plate.

Permittivity ϵ. The dielectric constant or permittivity ϵ_0 of a vacuum is 8.854×10^{-12} farad m^{-1} in the mks system of units.

Relative Dielectric Constant or Relative Permittivity ϵ_r. The factor by which the capacitance of a vacuum capacitor is increased by substitution of a dielectric for the vacuum.

Polarizability α. The proportionality constant between the induced dipole moment and the local electric field.

Polarization **P.**   The total dipole moment per unit volume in a dielectric.

Dielectric Strength (Breakdown).   The limiting voltage gradient required to cause appreciable current flow or failure of a dielectric.

Ferroelectric.   A material which exhibits spontaneous polarization, and electric hysteresis.

Electrostriction.   The change in dimensions of a dielectric due to an applied field.

Piezoelectric.   A crystal which becomes polarized when stressed.

Relaxation Time.   The time required for a disturbed system to reach $1/e$ of the final equilibrium configuration.

Relaxation Frequency.   The reciprocal of the relaxation time.

BIBLIOGRAPHY

SUPPLEMENTARY READING:

Cady, W. P., *Piezoelectricity,* McGraw-Hill, New York, 1946.

Dekker, A. J., *Electrical Engineering Materials,* Prentice Hall, Englewood Cliffs, N. J., 1959.

Kingery, W. D., *Introduction to Ceramics,* John Wiley & Sons, New York, 1960.

Megaw, A. D., *Ferroelectricity in Crystals,* Methuen & Co. Ltd, London, 1957.

Murphy, E. J., and Morgan, S. D., "Dielectric Properties of Insulation Materials," *Bell System Technical Journal,* **16**, 493 (1937); **17**, 640 (1938); **18**, 502 (1939).

Smyth, C. P., *Dielectric Behavior and Structure,* McGraw-Hill, New York, 1955.

Von Hippel, A., *Dielectric Materials and Applications,* M.I.T. Press and John Wiley & Sons, New York, 1956.

ADVANCED READING:

Debye, P., *Polar Molecules,* Dover Publications, New York, 1945.

Dekker, A. J., *Solid State Physics,* Prentice-Hall, Englewood Cliffs, N. J., 1957.

Kittel, C., *Introduction to Solid State Physics,* 2nd Edition, John Wiley & Sons, New York, 1956.

PROBLEMS

12.1   Given a parallel plate capacitor 10×25 cm, filled with a dielectric of relative permittivity 5:

(a) Calculate its capacitance.

(b) Calculate the charge on the plates.

(c) What is the value of D in the dielectric?

12.2   (a) Derive an expression for the time dependence of the current and voltage across a capacitor in series with a resistor, where only constant (d-c) voltages are applied to the series circuit.

(b) If the capacitance were 1 microfarad, the resistance 0.5 ohms and 200 volts suddenly applied, how long will it take for the voltage across the capacitor to rise to 100 volts?

12.3 *Leakage* through a capacitor by conduction of electrical charge through the dielectric can affect the time needed to charge and discharge, and the loss angle δ. Explain, using a parallel resistance to represent the leakage.

12.4 (a) In a resistor-capacitor series circuit containing a capacitor with ideal dielectric, derive an expression for the charging current if the impressed voltage varies sinusoidally.

(b) Show that the current through the capacitor is exactly 90° advanced in phase in relation to the applied voltage across the capacitor. How about the entire circuit?

12.5 (a) In dielectric subjected to a field of 10^{16} cps, what mechanism is responsible for the value of the relative permittivity?

(b) How is it related to the index of refraction?

(c) If the value of ϵ_r for a glass is 6.75 at frequencies 10^9 cps, what percentage may be attributed to ionic polarizability?

12.6 (a) The index of refraction of boron oxide is 4, but the electrically measured relative permittivity is 34. What is the percentage contribution of the ionic polarizability in this case?

(b) Why would you expect the relative permittivity 16 (obtained by squaring the index of refraction obtained from optical measurements in the infrared) to match the low frequency electrical measurements?

12.7 (a) Why should ion-jump polarization occur more easily at higher temperatures and result in a lower loss angle?

(b) Why should the dielectric strength of titania ceramics all tested in oil become lower, the higher the porosity?

12.8 (a) When alkali ions are added to fused quartz, the glass thus formed (see Volume I) has a higher conductivity and the loss factor increases. Explain in terms of the changes which occur in the glass network structure. (Remember the bonding of the silicate tetrahedra!)

(b) The Na^+ ion has greater mobility than K^+ or Rb^+ ions; would the loss factor of a silicate glass be affected if Rb^+ ions are substituted for some of the Na^+ ions?

(c) Why should the substitution of large divalent Ba^{2+} and Pb^{2+} ions for Na^+ ions in a glass favor both a lower melting point and a lower loss factor?

12.9 Why should the cooling of barium titanate ceramics through the Curie temperature in the presence of a strong electric field improve the piezoelectric properties of the product?

12.10 The coupling coefficient of a piezoelectric material indicates the fraction of the applied mechanical stress which is converted into an electrical voltage. Prepolarized barium titanate ceramics have a five times higher coupling coefficient than quartz and other useful piezoelectrics. What advantages or disadvantages have $BaTiO_3$ ceramics over other useful piezoelectric materials?

12.11   Demonstrate with a sketch how the perovskite structure of barium titanate can become polarized by motion of the entire anion lattice along a $<100>$ direction.

12.12   Regular oil changes are prescribed for oil-filled high voltage transformers. Why?

12.13   Derive Equation 12.8 for the dielectric power loss.

CHAPTER THIRTEEN

Optical Properties

Dielectrics are used to absorb, emit, and refract various types of radiation. When light passes through a dielectric, it can be reflected and refracted at the surface and absorbed and scattered within the material itself. Double refraction, or birefringence, occurs when the index of refraction is a function of crystal orientation. The index of refraction is a function of wavelength; this dependence is called optical dispersion. The selective emission and absorption of specific frequencies can be connected with the presence of impurities and imperfections which provide localized energy levels in the forbidden energy gap. Luminescent materials generally contain impurities called activators, which contribute energy levels which are responsible for the emission of visible light. The photographic decomposition of silver bromide by light absorption involves electrons, holes, and impurities. If the emission of radiation is stimulated rather than spontaneous, coherent monochromatic radiation may be emitted.

13.1 INTRODUCTION

The broad range of the electromagnetic spectrum is shown in Figure 13.1. The shortest wavelength radiation is emitted by radioactive materials and the longest (intentionally) by radio oscillators. Some radiant sources like incandescent solids emit a continuous range of wavelengths, others like arcs and sparks, descrete spectra. A beam of light from such sources is made up of many wave trains of different wavelengths with completely random phase relationship, because the electronic energy transitions responsible for the radiation occur independently. In con-

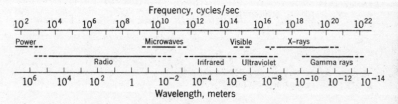

Figure 13.1   The electromagnetic spectrum, on a logarithmic scale.

trast, a radio beam from an antenna (dipole) is coherent, that is, its amplitude, phase and wavelength are controlled, and the electrons are forced to synchronize their radiation. A radio beam can, therefore, be used to carry detailed signals, whereas ordinary light beams are suitable only for crude, simple information, e.g., blinking. To obtain coherent light, it was necessary, until recently, to filter, polarize, and disperse ordinary light until a single, coherent, extremely weak component is left. Today, specially designed lasers emit light of very high coherency and intensity.

No solids reflect or transmit all of the incident radiation. All reflect and absorb some radiation. Most metals are good reflectors: some insulators, good transmitters while some semiconductors reflect in the blue, and transmit in the infrared. As a general rule, any good absorber is, at some frequency a good reflector.

Most of the important optical phenomena such as reflection, refraction, interference and polarization can be adequately described for the purpose of this chapter by the wave theory of light. Selective absorption and emission of light are better described by light quanta or photons.

13.2   REFRACTION

The index of refraction, n, of a material is defined as the ratio of the velocity of light in vacuum c to the velocity of light in the material v:

$$n = \frac{c}{v} \qquad (13.1)$$

The velocity of light in any medium is equal to $1/\sqrt{\epsilon\mu}$. Most transparent media are nonmagnetic and $\mu \simeq \mu_0$. Therefore, using Equation 13.1 the index of refraction is related to the relative permittivity by

$$n^2 = \frac{\epsilon\mu}{\epsilon_0\mu_0} \cong \frac{\epsilon}{\epsilon_0} = \epsilon_r \qquad (13.2)$$

Since the permittivity is a function of frequency, the velocity of propagation of light in a solid is also a function of frequency. This phenomenon is called *dispersion*. The permittivity of solids is relatively large due to the interaction of the dipoles in the solid with the electromagnetic radiation. Dense materials have more dipoles per unit volume, and therefore generally have high indices of refraction. Consequently, the addition of Pb or Ba to silica glass increases the density and ϵ_r and n. Figure 13.2 shows the refractive index plotted as a function of wavelength for three glasses of different density. The different forms of quartz all have the same chemical composition but a different density, depending

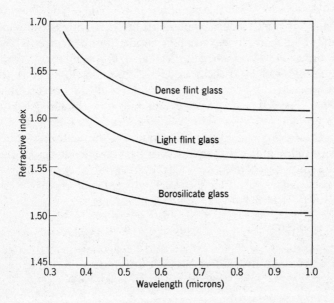

Figure 13.2   Refractive indices as a function of wavelength for three glasses. (After W. O. Kingery.)

Table 13.1

MATERIAL	AVERAGE REFRACTION INDEX	BIREFRINGENCE
Silica glass SiO$_2$	1.46	—
Soda-lime glass	1.51	—
Pyrex (borosilicate)	1.47	—
Flint glass	1.7	—
LiF	1.39	—
NaF	1.33	—
CaF$_2$	1.43	—
Al$_2$O$_3$	1.76	0.008
MgO	1.74	—
SiO$_2$ quarts	1.55	0.009
TiO$_2$	2.71	0.287
PbO	2.61	—
PbS	3.91	—

on their packing factors. Closely packed crystals of MgO and Al$_2$O$_3$ have a higher density and refractive index than SiO$_2$. PbS has an even higher index. The index of refraction is also greater in crystals whose atoms have higher atomic numbers, that is, more electrons. Thus sodium chloride has a higer index (1.54) than sodium fluoride (1.33) because the Cl$^-$ ions contribute more electrons than the F$^-$ ions. This makes the dielectric polarizability greater. The index of refraction of a number of materials is listed in Table 13.1.

In solutions or mixtures, the index of refraction is not an average of the indices of the separate components, since it depends on the sum of the dielectric polarizabilities. A sugar solution has the same index as a solid sugar crystal because the polarizability of sugar does not change in aqueous solutions.

The index of refraction of glass and most highly symmetric crystals (cubic) does not change with direction, but in crystals of lower symmetry, e.g., tetragonal and hexagonal crystals, it does. The latter have two different indices of refraction, while orthorhombic, monoclinic, and tegragonal crystals have three. The difference between the highest and lowest index is taken as a measure of the *birefringence* of the crystal (see Table 13.1). Birefringence depends on the bonding and packing in different crystallographic directions.   Cubic cristobalite, one of the crystal-lographic forms of quartz, has zero birefringence, whereas rhom-bohedral calcite ($CaCO_3$) has a large birefringence.   This is be-cause tightly bound triangular CO_3^{-2} groups are oriented normal to the c axis resulting in greatly different polarizability parallel to and normal to this axis.   Birefringence can also be induced in glass and polymers by the application of stresses which introduce asymmetric density changes.   Photoelastic analysis, which ena-bles stress distributions in transparent plastic models to be measured optically, is based on this effect.

An optically anisotropic or birefringent crystal is also doubly refracting, that is, it will split a beam of incident light into two beams, as shown in Figure 13.3.   One beam, the *ordinary beam,* emerges as a continuation of the incident beam.   The other, the *extraordinary beam,* is displaced from the incident beam.   The two beams are polarized, at right angles to each other.   Doubly refracting crystals can, therefore, be used to make polarizing

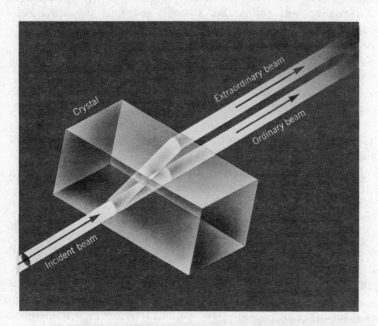

Figure 13.3   Double refraction by an anisotropic crystal.

Table 13.2   Indices for Some Common Birefringent Crystals

CRYSTAL	n ORDINARY	n EXTRAORDINARY
Calcite $CaCO_3$	1.66	1.49
Quartz SiO_2	1.54	1.55
Tourmaline	1.64	1.62
Ice H_2O	1.306	1.307

prisms, such as the Nicol prism, which consists of two halves of doubly refracting crystal cemented together. The cement (usually Canada Balsam) has an index of refraction such that the ordinary beam is totally reflected, and the extraordinary beam transmitted, that is, the angle of incidence of the ordinary beam is within the circle of total reflection. In other doubly refracting, polarizing crystals, one polarized component is absorbed much more strongly than the other. Such crystals are called dichroic. No prism need be constructed, as such materials polarize directly. Tourmaline and iodoquinone sulfate are strongly dichroic, the latter even more so when its polymeric chains are carefully oriented by straining (see Volume I). In such a condition, it is known commercially as Polaroid. Table 13.2 lists indices for some common doubly refracting crystals.

In mineralogy and glass technology, refractometers are used to identify optical phases by measurements of the refractive index, birefringence, and dispersion. They often provide more practical information than X-ray diffraction. Control of the processing of glass from melt to finished lens requires careful inspection of surface and interior for residual stresses as well as examination for obvious inclusions, bubbles and other flaws.

Light at normal incidence is partially reflected at the front surface of a dense optical medium, according to the relation:

$$R = \left[\frac{n-1}{n+1} \right]^2 \tag{13.3}$$

where R is the fraction of the incident intensity reflected, and is called the *reflectivity*. At normal incidence R varies from about 0.03 for low index glass (1.4) to about 0.08 for high index glass ($n = 1.8$). To cut down such losses, lenses are often coated with a thin layer (about one quarter wavelength) of isotropic transparent MgF_2, which reduces reflection by causing destructive interference of the light reflected from the glass-film interface. Multiple layers of thin dielectric films of alternating high and low indices of refraction have a high reflectivity due to constructive interference of incident light.

13.3 ABSORPTION

The fraction of incident light transmitted by a transparent optical material depends on losses due to absorption as well as reflection. For a given wavelength,

$$A_\lambda + R_\lambda + T_\lambda = 1 \qquad (13.4)$$

where A_λ is the fraction absorbed, R_λ the fraction reflected, and T_λ the fraction transmitted, using light of wavelength λ. Figure 13.4 indicates how the color of a green glass depends on these factors. The fraction absorbed depends on the wavelength as well as the optical path. Absorption as a function of path length is given by the relation:

$$\frac{dI}{I} = -\alpha \, dx \qquad (13.5)$$

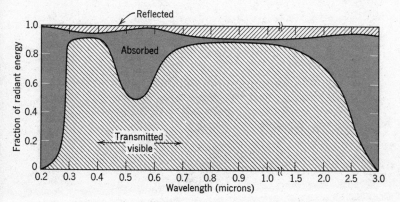

Figure 13.4   The fractions of incident light absorbed, transmitted and reflected by green glass, as a function of wavelength.

or

$$I = I_0 e^{-\alpha x} \tag{13.6}$$

where I_0 is the intensity of incident light, I that of the transmitted light, and α the absorption coefficient. We can apply these ideas to the transmission of light through a piece of material, as Figure 13.5 shows. The fraction T of incident intensity transmitted is, considering the reflections at both surfaces and the absorption,

$$T = (1 - R)^2 e^{-\alpha x} = \frac{I \text{ (transmitted)}}{I \text{ (incident)}} \tag{13.7}$$

The absorption is also dependent on the chemical composition of the absorber. The absorption or its converse, the transmission of glass can be varied in different regions of the spectrum by the addition of other ions. Pure silica glass will transmit visible and ultraviolet light down to about 2000 Å, whereas a borosilicate glass will only transmit down to 3000 Å; soda-lime glass used as windows transmits down to 3500 Å. Additions of FeO to ordinary soda-lime glass reduces transmission in the red end of the visible portion of the spectrum.

Total opacity, or translucency in solids, is usually due to extensive internal reflection and refraction. The opacity of wall plaster,

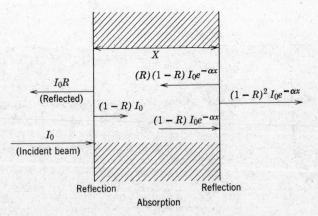

Figure 13.5 Transmission of light through a material in which reflections occur at both surfaces and absorption within.

enamels, and glazes, as well as the translucency of salts, silicates, and pigmented materials, all depend on multiple internal reflection and refraction. Structural inhomogeneities in materials (such as second phases) can act in the manner described in Figure 13.5, reflecting and absorbing, and they can also scatter the light. In general, fine dispersions of a second phase are more effective scatterers than coarse dispersions, for a given fraction of the second phase, because there are more particles to act as scatterers. The intensity as a function of path length is exponential,

$$I = I_0 e^{-sx} \tag{13.8}$$

where s is called the *scattering coefficient* of the solid.

13.4   ABSORPTION IN DIELECTRICS

Dielectrics such as glass, quartz, and polyethylene transmit visible light, and exhibit appreciable absorption in the near or far ultraviolet when relatively free of impurities or secondary phases. If impurities are naturally present or purposely added, selective absorption of visible light occurs. Such impurities introduce localized energy states (like acceptor and donor levels, but much higher) in the wide forbidden energy gap of insulators. Absorption occurs because incident photons of energy less than the gap width can for example excite electrons from the valence band up to the impurity states lying below the conduction band. Such absorption has been intensively studied in alkali-halide crystals.

When a transparent colorless halide crystal like sodium chloride is heated in a sodium vapor, it turns yellow. The color arises from imperfections called F centers or color centers. As sodium atoms deposit on the crystal from the vapor, chlorine ions migrate from the interior to combine with the sodium. This leaves ionic vacancies in the crystal interior. These act as traps for the extra electrons resulting from compound (NaCl) formation at the surface, which must remain in the crystal to preserve charge neutrality. The energies of the trapped electron levels lie in the forbidden energy gap. A crystal which has been prepared in this fashion can absorb photons of the proper frequency from a continuous spectrum of incident light to raise the trapped electrons to one of the higher energy states within the energy gap.   It is found that the total number of photons absorbed is directly proportional to the number of vacancies present in the crystal.   The absorption peak or *F band* observed in sodium-vapor treated sodium chloride crystals lies in the yellow, but can be shifted to somewhat longer wavelengths by heating and shorter wavelengths by cooling the crystal.   The absorption maximum also broadens on heating. Colored halide crystals can also be bleached by illuminating the crystal with light, which is of the same or close to the same wavelength as the absorption maximum.   Bleaching is attributed to the ionization of the F centers and is accompanied by photoconductivity.

Glasses can be colored by introducing transition or rare-earth metals when the glass is molten.   In ceramic materials, colorants are introduced in the bulk or in surface coatings.   In paints, pigment particles are not only chosen for their color but also for their high index of refraction as compared to the linseed oil carrier phase.   This insures multiple internal reflections which enhance the color.   Increasing pigment particle size in paints also increases the normal reflectivity at the expense of the diffuse reflectivity observed at other angles than the normal.   Objective measurement of the color of surfaces is not only difficult from an optical standpoint but also for physiological and psychological reasons.

13.5   PHOTOGRAPHIC IMAGES

The selective absorption characteristics of silver bromide are utilized in photographic film.   Sheets of cellulose acetate or glass plates are coated with a thin layer of gelatin containing fine grains of silver bromide (or another halide).   The silver bromide decomposes by absorbing light of the proper wavelength.   After exposure to light and partial decomposition of the silver bromide, the silver photographic image produced is developed in an alkaline solution of hydroquinone.   The fixing bath of sodium thiosulfate is then used to dissolve the remaining undeveloped grains of silver bromide.

The decomposition of the silver bromide is a complicated process, involving electrons, holes, sulfide impurities, and silver interstitials.   According to the model proposed by Gurney and Mott

(1938), an incident photon striking a crystallite of silver bromide in the emulsion produces an electron-hole pair, with the electron excited to the conduction band, as pictured in Figure 13.6. The

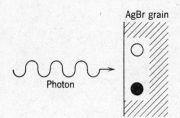

(*a*) A conduction electron and hole are created by an incident photon

(*b*) Surface trapping: the hole neutralizes a Br$^-$ ion, and the electron is trapped at the surface by Ag$_2$S or similar impurity

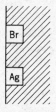

(*c*) An interstitial Ag$^+$ ion is attracted and neutralized by the trapped electron

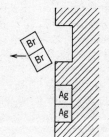

(*d*) Steps (*a*), (*b*), and (*c*) repeat. The Br atoms unite to form Br$_2$ and escape. The Ag nucleus grows into a tiny colloidal particle

Figure 13.6   First steps in formation of a photographic image.

electron migrates to the surface of the crystal where it becomes trapped, possibly by a silver sulfide (AgS) impurity. An interstitial silver ion is attracted from a Frenkel defect by the electron, and is neutralized. As more electrons are trapped, the process is repeated at the same site, and a silver nucleus grows. The holes neutralize the bromine ions, which then escape as bromine molecules. Further accumulation of silver atoms by the same process quickly leads to the formation of a colloidal precipitate called a *latent image*. The hydroquinone (or other chemical) developer continues the growth of the silver particle, until the entire AgBr grain is converted. The fixer makes the remaining underdeveloped grains insensitive to light, by removing the undecomposed silver halide. Silver bromide emulsions are, generally, more responsive to blue light than red. The color response can be made more even by special dyes which increase the range of colors to which the crystal responds. The film is then called *panchromatic*.

13.6 LUMINESCENCE

Many inorganic and some organic substances can absorb energy and emit visible or near-visible radiation. Such behavior, called *luminescence,* can be initiated by bombardment with photons, electrons, or positive ions, and by mechanical stress, chemical reaction, or heating. In industrially important luminescent materials, absorption of ultraviolet light is followed by emission of visible light, e.g., in fluorescent lamp tubes. If the emission takes place within 10^{-8} seconds of excitation, a luminescent material is referred to as *fluorescent.* If the emission takes longer, the material is called *phosphorescent.* Typical *phosphors* include sulfides, silicates, oxides, tungstates, and organic materials such as anthracene. Controlled amounts of impurities are put into phosphors to provide localized states in the energy gap; these impurities are called *activators.*

The insides of fluorescent lamps are coated with silicates or tungstates, which are excited by the ultraviolet light from a mercury glow discharge. Such lamps can be five times as efficient as standard incandescent bulbs. To increase the "whiteness" of luminescent emission in fluorescent lamps, two kinds of activators

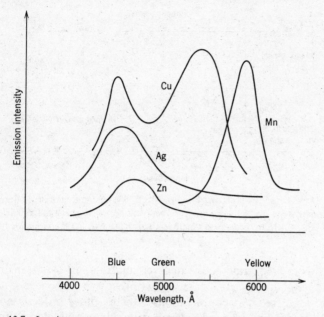

Figure 13.7 Luminescence spectra for ZnS phosphors with ultraviolet excitation. (After H. W. Leverenz, *Introduction to Luminescence in Solids,* Wiley, 1950.) The curves labeled Cu, Ag, and Mn are for ZnS activated with those metals. The curve labeled Zn is for "self-activated" ZnS, in which activation is accomplished by zinc vacancies associated with impurity (donor) levels, notably halogens.

are used, one for blue and the other for yellow orange emission. The effect of different activators on the color and intensity of emission of ZnS phosphors is shown in Figure 13.7. The copper ions, introduced as activators, replaces Zn atoms in the lattice of ZnS and are believed responsible for the maximum shown in the green, while the blue emission is attributed to vacancies which arise as the result of singly ionized atoms of silver or copper. In CdS and other substances, off-stoichiometric compositions lead to self-activation. This results from the presence of inherent vacancies with energy levels in the forbidden gap. Organics like anthracene can be activated with naphtacene.

An effective activator level is one which the electron finds it easy to enter and leave. Otherwise the electron may prefer to recombine directly, by descending to the valence band. Figure 13.8 shows two possibilities for recombination. In one, the electron

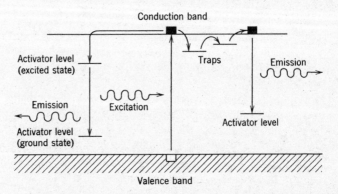

Figure 13.8   Activator levels and luminescence. The emission may occur during a transition between two activator levels, or between the conduction band and an activator level. An electron can be thermally excited from a trap to the conduction band.

descends to an excited activator level, and radiates by a transition to the activator ground state. This process is infrequent at all except high free carrier transitions. In the other process shown the electron is first trapped in levels which do not allow transitions involving radiation. The electron is eventually thermally excited to the conduction band, and then radiates by going to an activator level.

Insulator panels coated with an appropriate phosphor and connected to a source of electricity can also be made luminescent. Electroluminescence of this kind has been observed in CdS, SiC, Ge, Si, and several III-V type compounds.

13.7   LASERS

The light emitted by fluorescent lamps or other light sources results from energy transitions from higher states to lower ones. Atoms of the same element make similar transitions and therefore give off photons of similar wavelengths. They do so, however, independent of one another and at random times. This results in the emission of incoherent radiation because it is out of phase.

Since the early work of Weber (1952) and Townes (1955) on stimulated emission of electromagnetic radiation, solid-state and gas-light sources have been developed which give coherent radiation. Such light sources are called *lasers,* which is an abbreviation for *light amplification by stimulated emission of radiation.* To describe the operation of such a device the solid state ruby laser introduced by Maimann in 1960 provides a convenient example.

The ruby laser shown in Figure 13.9 is a single crystal of corundum (Al_2O_3) doped with 0.05 percent Cr^{3+} ion. The chromium ion gives the ruby its characteristic pink color. It also provides the active fluorescent centers which are responsible for the characteristic emission. The crystal is first machined and the ends ground parallel to optical tolerances. The ends are then silvered, one opaque and one slightly transparent. Light of high intensity from the xenon flash lamp provides, by its component green radiation, the energy necessary for excitation of the Cr^{3+} ions of the ruby. As a result of the fluorescent decay of such excitation the laser emits a highly collimated beam of coherent radiation whose wavelength lies in the red (6943 Å).

Before the xenon lamp is flashed, the Cr^{3+} ions are essentially all in the ground state as indicated schematically by the open

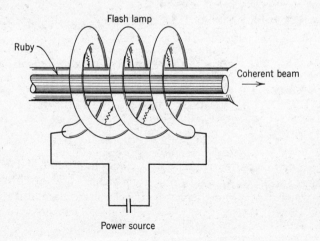

Figure 13.9   Schematic of ruby laser.

circles on Figure 13.10a. Photons of wavelength 5600 Å emitted (among others) by the lamp raises the Cr^{3+} ions to high lying energy states. Such excited ions can either drop back to the ground state or to the metastable state characteristic of the Cr^{3+} ion. The shaded circles of Figure 13.10b represent the latter state in which the ions can remain for up to 3 milliseconds before spontaneous emission occurs. The first few spontaneously emitted photons can however stimulate the remaining Cr ions

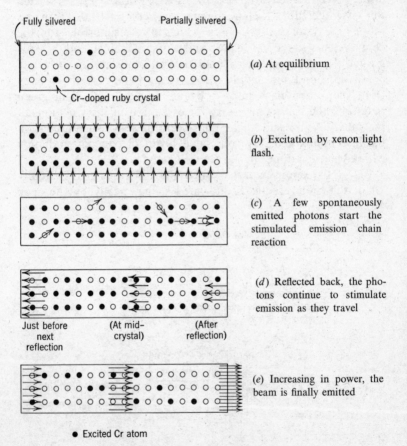

(a) At equilibrium

(b) Excitation by xenon light flash.

(c) A few spontaneously emitted photons start the stimulated emission chain reaction

(d) Reflected back, the photons continue to stimulate emission as they travel

(e) Increasing in power, the beam is finally emitted

● Excited Cr atom

○ Cr atom in ground state

Figure 13.10 Schematic showing stimulated emission and light amplification in ruby laser.

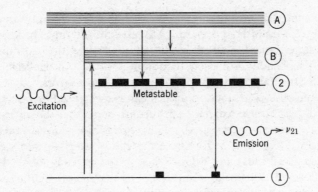

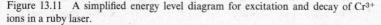

Figure 13.11   A simplified energy level diagram for excitation and decay of Cr^{3+} ions in a ruby laser.

lying in the metastable state to emit and initiate the chain reaction represented in Figure 13.10d.   The photons emitted as the result of such stimulation are in phase with the stimulating photons and have the same frequency.

The narrow beam divergence and high intensity of the emitted radiation is promoted by back and forth reflection between the end mirrors as indicated in Figure 13.10d.   Photons emitted at an angle to the axis of the crystal are lost.   The high intensity pulse finally emitted through the partially silvered end of the laser as indicated in Figure 13.10e is of relatively short duration (0.6 msec). This type of stimulated radiation in the modern ruby laser is more than 10^4 times as great as the intensity of natural ruby fluorescence.   Other solid state lasers such as CaF_2 (doped with 0.05% Nd ion) and neodynmium glass are used to obtain pulsed radiation of longer wavelengths.

The energy transitions of the Cr ions in the ruby laser are shown in the simplified energy diagram of Figure 13.11.   The arrows pointing upward represent excitation and those downward, decay. The transition from level 2 to level 1 is the only one that leads to stimulated emission and coherent radiation.   (See Problem 13.14.)

Solid state lasers can not be operated for long periods because of the heat generated.   Although gas lasers have lower intensity they are more suitable for some purposes since they can be operated continuously.   In one example of a gas laser the high frequency electrical discharge tube used contains He at 1 mm and Ne at 0.1 mm of Hg pressure. Atoms of the latter gas produce the stimulated emission which lies in the infrared. Other gas mixtures and proportions are used to obtain shorter wavelengths. Gas lasers are equipped with special external end mirrors.

Semiconductor junction diodes can also be used as lasers, provided the minority carriers can recombine so that the energy lost by the electrons is radiated. This is possible in GaAs junctions, but not in Ge and Si for there the energy is dissipated as heat. Figure 13.12 illustrates a GaAs laser with optically finished sides which serve as mirrors. Conversion of electrical to light energy in this laser is relatively efficient and the output can be controlled by the voltage. At sufficiently high voltages and recombination rates, photon simulated recombination and emission will occur. By doping with phosphorus wavelengths in the range 6500 to 8400 Å have been obtained. Longer wavelengths can be obtained with InAs lasers.

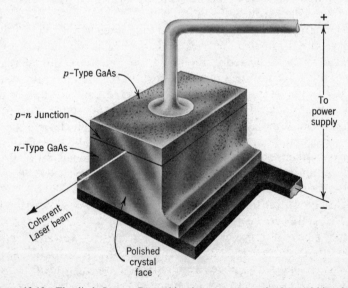

Figure 13.12   The diode Laser.   Recombination occurs near the forward-biased p-n junction; the electrons radiate as they descend to the valence band. This transition can be stimulated by photons, thus providing the Laser action.

Light beams from lasers have potential applications in many areas. Because they are coherent, they can be used to carry signals by amplitude modulation. Hence development for communication purposes is active. The high intensity and monochromaticity of laser light also makes it useful in photographic and spectrographic research. Since the laser beam can be easily focussed, it has been used for particular surgical and metallurgical purposes where highly localized heating is mandatory. Narrow beam divergence and high intensity also make the laser beam useful in precision measurement and surveying.

DEFINITIONS

Monochromatic Light.   Light of a single wavelength.

Coherent Radiation.   Cooperative and single phase radiation.

Index of Refraction (n).   The ratio of the velocity of light in vacuum to that in the material of interest.

Activators.   Impurities added to insulators to produce impurity levels in the energy gap, which aids in the production of luminescent light.

Laser.   Abbreviation of "light amplification by stimulated emission of radiation." A light source which gives coherent radiation.

Spontaneous Emission.   Radiation due to the decay of an excited state which has not been perturbed after excitation.

Stimulated Emission.   Radiation due to the decay of an excited state, which has been induced to radiate by a photon of the frequency corresponding to the decay emission.

Optical Dispersion.   The variation in index of refraction of an optical medium with wavelength.

Absorptivity.   The fraction of incident light absorbed by a material.

Color Center (F Center).   Electron bound to a negative ion vacancy which introduces additional energy levels in the forbidden energy gap of an insulator.

Luminescence.   Absorption of light or other energy by a substance followed by spontaneous emission of light of longer wavelength.

Fluorescence.   Light emitted simultaneously or within 10^{-8} sec of its excitation.

Phosphorescence.   Light emitted from a phosphor at times more than 10^{-8} sec after excitation.

Electroluminescence.   Light given off by a phosphor when excited by an electric field.

BIBLIOGRAPHY

SUPPLEMENTARY READING:

Feynman, R., *The Feynman Lectures on Physics,* Addison-Wesley, Reading, Mass., 1964.

Kerr, P. F., *Optical Minerollogy,* 3rd Edition, McGraw-Hill, New York, 1959.

Kingery, W. D., *Introduction to Ceramics,* Wiley, New York, 1960.

Leverenz, H. W., *An Introduction to Luminescence in Solids,* New York, 1950.

Sears, F. W., *Optics,* Addison-Wesley, Reading, Mass., 1949.

Weber, S., Ed., *Optoelectronics Devices and Circuits,* McGraw-Hill, New York, 1964.

ADVANCED READING:

Dekker, J. A., *Solid State Physics,* Prentice-Hall, New York, 1957.

Kittel, C., *Introduction to Solid State Physics,* 2nd Edition, Wiley, New York, 1956.

Mott, N. F., and Gurney, R. W., *Electron Processes in Some Crystals,* 2nd Edition, Oxford University Press, London, 1957.

PROBLEMS

13.1 (a) Most glasses and ionic crystals are transparent to visible light, but opaque to ultraviolet. Explain.

(b) Some ionic crystals are opaque to visible light. Fe_3O_4, $FeTiO_3$, and FeS_2 are examples. Explain.

13.2 "Coated" lenses have a dielectric film, one quarter wavelength thick, on the surface, which drastically reduces the reflectance. Explain why this occurs.

13.3 Extremely reflective mirrors can be constructed of thin layers of dielectric films. Two substances are alternated in the layers, so that layers of extremely high refractive index are alternated with layers of extremely low refractive index. Explain how such a mirror works. (*Hint.* Use equation 13.3.)

13.4 (a) Colored alkali halides are not as dense as the transparent versions of the same halides. Why?

(b) Explain how colors produced by F centers can be bleached by light of the same color.

(c) As a hint for part *b*, and a third question. During bleaching by light, colored alkali halide crystals become conductive, that is, they are photo-conductive—why?

13.5 Compare optical absorption with the photoelectric effect. What are the similarities, and what are the differences? Consider also stimulated emission.

13.6 Describe an experiment that you could readily do in the laboratory to show the difference between coherent and incoherent radiation.

13.7   The emission wavelength from a GaAs Laser can be changed from 9000 to about 7000 A.U. by substituting phosphorus for some of the arsenic. Explain.

13.8   In order to send AM (amplitude modulated) signals, the power output of a Laser must be modulated, e.g., by an audio frequency signal. How could this be accomplished, for ruby and for GaAs Lasers? What are the implications of your conclusions?

13.9   Gem stones can be colored by exposure to neutron irradiation. (It has been proposed, for instance, that yellow diamonds, which are relatively cheap, be converted to blue or blue-white stones by irradiation.) Can you explain how the coloring might occur?

13.10   High-speed photographic film sometimes suffers from excessive grain size, especially when it has been overdeveloped; consequently, resolution is poor. Explain both the effect of high speed and overdevelopment.

13.11   (Library Problem) Explain the operation and possible applications of "fiber optics." (*Reference. Optoelectronic Devices and Circuits* by S. Weber, McGraw-Hill, New York, 1964.)

13.12   Moving pictures have been taken at the rate of 10^8 frames per second, by using the Kerr cell, which consists essentially of a 100 to 1000 megacycle signal generator, a polarizing filter, and a tank of liquid which becomes birefringent when an electrical field is applied. Explain how the Kerr cell acts as a shutter.

13.13   State as briefly as possible what is meant by the term *metastable energy level.*

13.14   The metastable level of the ruby laser described in Section 13.7 is actually a double level. How does this affect the wavelength and the coherency of the stimulated emission? See Maimann, T. H., *Nature,* Vol. 187, p. 493 (1960).

13.15   In order to achieve *lasing,* a so-called population inversion is required in which the number of atoms in the excited state exceeds those in the ground state. Explain why this is necessary to achieve intense stimulated emission in a laser. See *Optoelectronic Devices and Circuits,* Ed. S. Weber, McGraw-Hill, New York, 1964.

13.16   With the aid of a simplified energy diagram, describe how a gas laser functions. (See the reference quoted in Problem 13.14.)

13.17   (a) Define coherent electromagnetic radiation.

(b) Distinguish between time and space coherency.

(c) With what other devices than lasers can one obtain coherent radiation?

(d) Describe a simple experiment using a screen with two adjacent holes to prove that laser light is coherent.

Appendix A

QUANTITY	MKS UNIT	DEFINING EQUATION MKS	GAUSSIAN	GAUSSIAN UNIT	EQUIVALENCE
Mass, length, time	kilogram, meter, second			gram, centimeter, second	1 kg = 10^3 gm 1 m = 10^2 cm
Force	newton = kilogram-meter/second^2	$F = ma$		dyne = gram-centimeter/second^2	1 newton = 10^5 dyne
Work, energy	joule = newton-meter	$W = \int \mathbf{F} \cdot \mathbf{ds}$		erg = dyne-centimeter	1 joule = 10^7 erg
Power	watt = joule/second	$P = dW/dt$		erg/second	1 watt = 10^7 erg/sec
Charge	coulomb = ampere-second	$q = \int I\,dt$	$F = qq'/r^2$	statcoulomb; [statcoulomb] = [dyne]$^{1/2}$ [centimeter]	1 coulomb = $(c/10)$ statcoulomb $\simeq 3 \times 10^9$ statcoulomb
Current	ampere; [ampere] = [newton]$^{1/2}/\mu_0^{1/2}$	$I = dq/dt$		statampere = statcoulomb/second = $(1/c)$ abampere	1 amp = $(1/10)$ abamp = $(c/10)$ statamp $\simeq 3 \times 10^9$ statamp
Current density	ampere/meter^2	$J = \lim \Delta I/\Delta S$		statampere/centimeter^2	1 amp/m^2 = $(c/10^5)$ statamp/cm$^2 \simeq 3 \times 10^5$ statamp/cm^2
Electric field intensity	newton/coulomb = volt/meter	$\mathcal{E} = \lim \Delta\mathbf{F}/\Delta q$		dyne/statcoulomb = statvolt/centimeter	1 newton/coulomb = $(10^6/c)$ dynes/statcoulomb $\simeq (10^{-4}/3)$ dynes/statcoulomb
Electric field flux	newton meter^2/coulomb	$\Phi_e = \int \mathcal{E} \cdot \mathbf{dS}$		dyne-centimeter^2/statcoulomb; $[\Phi_e]$ = [statcoulomb]	1 newton-m^2/coulomb = $(10^{10}/c)$ dyne-cm^2/statcoulomb $\simeq (1/3)$ dyne-cm^2/statcoulomb

294

Quantity	MKS unit	MKS formula	Gaussian formula	Gaussian unit	Conversion
Capacitance	farad = coulomb/volt; [farad] = ϵ_0[meter]	$C = Q/V$		statfarad = statcoulomb/statvolt; [statfarad] = [centimeter]	1 farad = $(10^{-9}/c^2)$ statfarad $\simeq 9 \times 10^{11}$ statfarad
Electric dipole moment	meter-coulomb	$p_e = qr$		centimeter-statcoulomb	1 m-coulomb = $(10c)$ cm-statcoulomb $\simeq 3 \times 10^{11}$ cm-statcoulomb
Electric polarization	coulomb/meter^2	$P = \dfrac{\Sigma p}{V}$ $P = \epsilon_0 \mathscr{E}(\epsilon_r - 1)$		statcoulomb/centimeter^2	1 coulomb/m^2 = $(10^{-5}c)$ statcoulomb/cm^2 $\simeq 3 \times 10^5$ statcoulomb/cm^2
Relative permittivity or dielectric constant; electric susceptibility	dimensionless ratio	$\epsilon_r = D/(\epsilon_0 \mathscr{E})$ $= \epsilon/\epsilon_0$ $\chi = \epsilon_r - 1$	$\epsilon_r = D/\mathscr{E}$ $\chi = \dfrac{(\epsilon_r - 1)}{4\pi}$	dimensionless ratio (statcoulomb^2/dyne-cm)	κ same in both systems; *1 unit of χ (MKS) = $(1/4\pi)$ unit of χ (Gaussian)
Displacement	coulomb/meter^2	$D = \epsilon_0 \mathscr{E} + P$	$D = \mathscr{E} + 4\pi P$	dyne/statcoulomb	*1 coulomb/m^2 = $(4\pi \times 10^{-5}c)$dyne/statcoulomb $\simeq 12\pi \times 10^5$ dyne/statcoulomb of **D**
Resistance	ohm = volt/ampere [ohm] = [second/meter]/ϵ_0 = μ_0[meter/second]	$R = V/I$		statohm = statvolt/statampere; [statohm] = [second/centimeter]	1 ohm = $(10^9/c^2)$ statohm $\simeq (10^{-11}/9)$ statohm
Conductance	mho = ampere/volt	$G = I/V$		statmho = statampere/statvolt	1 statmho = $(c^2/10^9)$ statmho $\simeq 9 \times 10^{11}$ statmho
Resistivity	meter-ohm	$\rho = \mathscr{E}/J$		statohm-centimeter	1 m-ohm = 100 ohm-cm = $(10^{11}/c^2)$ statohm-cm $\simeq (10^{-9}/9)$ statohm-cm

QUANTITY	MKS UNIT	DEFINING EQUATION MKS	DEFINING EQUATION GAUSSIAN	GAUSSIAN UNIT	EQUIVALENCE
Electrical conductivity	mho/meter	$\sigma_e = J/\mathscr{E}$		statmho/centimeter	1 mho/m = $(c^2/10^{11})$ statmhos/cm $\simeq 9 \times 10^9$ statmhos/cm
Magnetic induction ($\mathbf{B}$)	weber/meter^2 = tesla = newton/ampere-meter	$\mathbf{F} = q(\mathbf{v} \times \mathbf{B})$	$\mathbf{F} = (q/c)(\mathbf{v} \times \mathbf{B})$	gauss = (c) dyne/stat-ampere-centimeter (= oersted); [gauss] = μ_0[oersted] = $\mu_0^{1/2}$[dyne$^{1/2}$/cm]	1 weber/m^2 = 10^4 gauss
Magnetic flux	weber = tesla-meter^2 = volt-second = joule/ampere = ampere-henry; [weber] = $\mu_0^{1/2}$[newton$^{1/2}$-meter]	$\Phi = \int \mathbf{B} \cdot \mathbf{dS}$		maxwell = gauss-centi-meter^2 = (c) erg/stat-ampere (= oersted/centimeter^2)	1 weber = 10^8 maxwells
Relative permeability μ_r; magnetic susceptibility χ_m	dimensionless ratio	$\mu_r = \mu/\mu_0$ $\chi_m = \mu_r - 1$	$\mu_r = \mu$ $\chi_m = \dfrac{\mu - 1}{4\pi}$	dimensionless ratio	μ_r same in both systems; *1 unit of χ_m (MKS) = $(1/4\pi)$ unit of χ_m (Gaussian)
Magnetic moment	$\begin{cases} \text{ampere-meter}^2 \\ \text{weber-meter} \end{cases}$	$\mathbf{m} = I\mathbf{A}$ $\mathbf{p}_m = \mu_0 I\mathbf{A}$	$\mathbf{m} = (I/c)\mathbf{A}$	gauss-centimeter^3 = erg/oersted (= erg/gauss)	1 amp-m^2 = $4\pi \times 10^{-7}$ weber-m = 10^3 gauss-cm^3 of $\mathbf{m}$
Magnetization	$\begin{cases} \text{ampere/meter} \\ \text{weber/meter}^2 \end{cases}$	$\mathbf{M} = \dfrac{\Sigma \mathbf{m}}{V}$		gauss (= oersted)	1 amp/m = $4\pi \times 10^{-7}$ weber/m^2 = 10^{-3} gauss (= 10^{-3} oersted) of $\mathbf{M}$

Magnetic intensity (**H**)	ampere/meter	$\mathbf{H} = \mathbf{B}/\mu_0 - \mathbf{M}$	$\mathbf{H} = \mathbf{B} - 4\pi\mathbf{M}$	oersted ($=$ gauss) $=$ $(1/c)$ statampere/centimeter; [oersted] $=$ [dyne$^{1/2}$/centimeter]	1 amp/m $= 4\pi \times 10^{-3}$ oersted of **H**
Hall coefficient	volt-meter³/ampere-weber $=$ meter³/coulomb	$R_H = \mathscr{E}_H/JB$		statvolt-centimeter/statampere-gauss $=$ centimeter³/(c) statcoulomb	1 v-m³/amp-weber $=$ $(10^7/c^2)$ statv-cm/statamp-gauss $\simeq (10^{-13}/9)$ statv-cm/statamp-gauss
Thermopower, Thomson coefficient	volt/degree Centigrade $=$ joule/coulomb-degree Centigrade	$\left. \begin{aligned} \dfrac{dV}{dT} &= S \\ \mu_{Th} &= \dfrac{TdS}{dT} \end{aligned} \right\}$		statvolt/degree Centigrade	1 V/°C $= (10^8/c)$ statv/°C $\simeq (1/300)$ statv/°C

Index